RURAL CHANGE IN AUS

T0228257

Perspectives on Rural Policy and Planning

Series Editors:
Andrew Gilg, University of Exeter and University of Gloucestershire, UK
Henry Buller, University of Exeter, UK
Owen Furuseth, University of North Carolina, USA
Mark Lapping, University of South Maine, USA

Other titles in the series

Keeping it in the Family
International Perspectives on Succession and
Retirement on Family Farms
Edited by Matt Lobley, John Baker and Ian Whitehead
ISBN 978 1 4094 0995 3

Participatory Rural Planning
Exploring Evidence from Ireland
Michael Murray
ISBN 978 0 7546 7737 6

Naming Food After Places
Food Relocalisation and Knowledge Dynamics
in Rural Development
Edited by Maria Fonte and Apostolos G. Papadopoulos
ISBN 978 0 7546 7718 5

A Living Countryside?
The Politics of Sustainable Development in Rural Ireland
Edited by John McDonagh, Tony Varley and Sally Shortall
ISBN 978 0 7546 4669 3

Rural Sustainable Development in the Knowledge Society
Edited by Karl Bruckmeier and Hilary Tovey
ISBN 978 0 7546 7425 2

Rural Change in Australia

Population, Economy, Environment

Edited by

RAE DUFTY-JONES
University of Western Sydney, Australia

JOHN CONNELL
University of Sydney, Australia

LONDON AND NEW YORK

First published 2014 by Ashgate Publishing

2 Park Square, Milton Park, Abingdon, Oxon OX14 4RN
711 Third Avenue, New York, NY 10017, USA

Routledge is an imprint of the Taylor & Francis Group, an informa business

British Library Cataloguing in Publication Data
A catalogue record for this book is available from the British Library

The Library of Congress has cataloged the printed edition as follows:
Dufty-Jones, Rae.
   Rural change in Australia : population, economy, environment / by Rae Dufty-Jones and John Connell.
      pages cm. -- (Perspectives on rural policy and planning)
   Includes bibliographical references and index.
   ISBN 978-1-4094-5204-1 (hardback : alk. paper)
   1. Regional planning--Australia. 2. Migration, Internal--Australia. 3. Australia--Economic policy--21st century. 4. Australia--Social policy--21st century. 5. Social change--Australia. I. Title.
HT395.A93D84 2014
307.1'2094--dc23
                                                                    2013034165

ISBN 978-1-4094-5204-1 (hbk)
ISBN 978-1-138-26709-1 (pbk)

Contents

Figures

Tables

Contributors

Neil Argent, Division of Geography and Planning, University of New England: nargent@une.edu.au

Louise E. Askew, Centre for Urban and Regional Studies, School of Environmental and Life Sciences, The University of Newcastle: laskew80@hotmail.com

Kerry Carrington, School of Justice, Faculty of Law, Queensland University of Technology: kerry.carrington@qut.edu.au

John Connell, School of Geosciences, University of Sydney: john.connell@sydney.edu.au

Amanda Davies, School of Built Environment, Curtin University: A.davies@curtin.edu.au

Danielle Drozdzewski, School of Biological, Earth and Environmental Sciences, University of New South Wales: danielled@unsw.edu.au

Rae Dufty-Jones, School of Social Sciences and Psychology, University of Western Sydney: r.dufty-jones@uws.edu.au

Chris Gibson, Australian Centre for Cultural Environmental Research, University of Wollongong: cgibson@uow.edu.au

Graeme Hugo, Australian Population and Migration Research Centre, The University of Adelaide: graeme.hugo@adelaide.edu.au

Pauline McGuirk, Centre for Urban and Regional Studies, School of Environmental and Life Sciences, The University of Newcastle: Pauline.mcguirk@newcastle.edu.au

Alison McIntosh, School of Justice, Faculty of Law, Queensland University of Technology: alison.mcintosh@qut.edu.au

Fiona McKenzie, Faculty of Humanities and Social Sciences, La Trobe University: fiona@australianfutures.org

Phil McManus, School of Geosciences, University of Sydney: phil.mcmanus@
sydney.edu.au

Bill Pritchard, School of Geosciences, University of Sydney: bill.pritchard@
sydney.edu.au

Fran Rolley, Division of Geography and Planning, University of New England:
frolley@une.edu.au

Meg Sherval, Centre for Urban and Regional Studies, School of Environmental
and Life Sciences, The University of Newcastle: meg.sherval@newcastle.edu.au

Erin F. Smith, School of Geosciences, University of Sydney: erin.smith@sydney.
edu.au

John Taylor, Research School of Social Sciences, Australian National University:
j.taylor@anu.edu.au

Jim Walmsley, Division of Geography and Planning, University of New England:
dwalmsle@une.edu.au

Preface

This book aims to provide a contemporary perspective on rapidly evolving population, economic and environmental changes in 'rural and regional Australia', itself a significant concept. Bringing together empirical and theoretical studies, it builds on established rural studies themes such as population change, economic restructuring and globalisation in agriculture but also brings into relief developments in this area around environmental change, culture, class, gender, and ethnic diversity in rural Australia. The book primarily seeks to present original and in-depth interventions on these issues and their intersections. While rural Australia is a vast area the book has sought to balance regional differences and interests, and to assemble the best of recent research on rural Australia. Such an approach provides important new material on contemporary population, economic and environmental changes in rural Australia, at a time when regional Australia is probably changing as fast as it has ever done. Mining is booming in some states, and wide-ranging debates have focused on issues of inter-state and international migration, the decline of some but not all country towns, the changing structure of agriculture, as farmers age, commodity prices fluctuate and environmental concerns mount, not least centred on the huge Murray-Darling basin. Access to water and climate change underpins many debates and complicates policy formation. Meanwhile rural Australia moves towards what has been called a 'post-productivist countryside' where production – of agricultural goods – has slowly given way to a different kind of Australia where creativity has contributed to new economic opportunities and new images of 'the bush'. While all of the chapters present empirical material they are also grounded in recent theoretical developments in human geography and rural studies concerning culture, nature, economics, identity and diversity.

We would like to thank the many contributors who responded with alacrity and enthusiasm to this project, in their own recognition of just how significant contemporary changes are for the future of Australia, and how the book might contribute to important ongoing debates. We would also like to acknowledge the role of the Institute of Australian Geographers in supporting the initial stages of this project. Last we would like to thank Kathryn Maiden, Erin Smith and Tim Jones for providing editorial and childcare support in the final stages of this book.

Rae Dufty-Jones and John Connell

Chapter 1

Introduction:
Twenty-first Century Rural Australia

John Connell and Rae Dufty-Jones

For only the second time in 100 years, the results from the 2010 Australian federal election produced a hung parliament. Among the five members of parliament who decided which political party would form the government for the next three years were three independents from rural electorates. While the role that these three men played in deciding the outcome of the 2010 election was a quirk of political fate, the election result itself raised a broad set of concerns about the impacts of a number of changes in which rural and regional Australia play a central role. Just as a series of electoral surprises in the closing years of the 1990s concentrated the minds of the nation's politicians on the anger and frustration of rural and regional Australia, the 2010 election once again alerted Australia's political representatives to the social, economic and political realities of a rural-urban divide that has been a characteristic and distinctive feature of much of Australia's history since white colonisation. In the agreement made between the Labor Party and two of the rural independents, Tony Windsor and Rob Oakeshott, rural and regional Australians were given front-row seats. One of the three substantial undertakings promised in this agreement was to restructure the 'government, public service and parliament to better meet the needs of regional Australia' (Australian Labour Party, Windsor and Oakeshott, 2010: 3). Implicitly development was disappointing and policies had failed. The document revealed both continuities and important differences in the demographic, economic and environmental factors used to justify this most recent political focus on rural and regional Australia.

An eighteen page annex that detailed how the 'Commitment to Regional Australia' should be delivered opened by noting that 30 percent of the nation's people lived outside the capital cities, continuing a well-worn argument that, while fewer than Australia's metropolitan populations, people living in the 'bush' should not be ignored. Yet glaring omissions persisted, with no mention of the contemporary role of immigrants in regional communities, or recognition of the multiple parts played (historical, economic, cultural and environmental) by Indigenous peoples in these landscapes. However, the population case differed from earlier iterations in the way it was nuanced through a narrowing of the focus onto the pressures faced by coastal communities and larger regional centres and the service needs of an ageing rural population.

Similar continuities could also be found in the case made around the economy. The theme of 'you need us [the rural] more than we need you [the city]', which prevailed when the Australian economy 'rode on the sheep's back', was contemporised in terms of food security. The document pointed out that 'the fresh food that ... is on the table of Australian families in cities, suburbs and towns across the nation' was a result of the labours of those in rural Australia (Regional Agreement 2010: 16). As in previous decades, agriculture and mining were celebrated as key sources of national export income, as well as important sources of scientific and industrial innovation. It was argued that it was in Australia's national economic interest to ensure that these key markets remain internationally competitive through the provision of necessary infrastructure. However, this document also staked a new economic claim on rural and regional Australia's behalf: for a 'fair share' of the economic opportunities presented by the digital economy. This share was to be provided by ensuring that 'every community in regional Australia gets fair and equal access to affordable high-speed broadband through the National Broadband Network' (Regional Agreement 2010: 20). Connectivity counts.

Last, the environmental case for why rural Australia is important also blended the old with the new. The environment was presented in the hackneyed colonial understandings of domestication: something that rural and regional Australians regularly battled and tamed. Discourses prevail in the national media around 'fighting' bushfires, 'battling' drought and 'surviving' floods and cyclones. The environment was also something to be secured against an imported other, whether it was in the form of 'biosecurity' and the protection of agriculture from foreign pests and diseases, or 'food security' and who owns and is responsible for the sources of national food production. Yet new discourses of 'protection' also influence how the environment was positioned in relation to rural and regional Australia. Coastal communities were identified as in need of protection from the consequences of climate change, and agricultural environments were positioned as being in need of protection from the negative of impacts of residential and mining development. Water straddles both understandings of the environment for rural and regional Australia, as an important resource needed for the taming of dry landscapes as well as being crucial for the protection of both river systems.

A Changing Rural Australia: What's New?

The theme of continuity and change, illustrated so well in the political negotiations of 2010, is foundational to this book's analysis of the rapidly evolving population, economic and environmental changes in rural and regional Australia. Change is far from a novel concept when it comes to rural studies, particularly those focusing on Australia. As Tonts *et al*. (2012: 291) observe 'an enduring concern within Australian rural geography [and the wider social sciences] has been to understand the nature and implications of change'. They further argued that change is so significant that it is fundamental in how we now understand rural Australia.

The 1990s and early-2000s saw a burgeoning of research that investigated the extent and impact of the changes that had occurred in rural and regional Australia since the 1970s. This work included texts such as Gray's (1991) ethnographic insights into the social and political workings of a country town in rural NSW and Alston's (1995) detailing of the role of women in negotiating, managing and, sometimes, instigating change. This period also witnessed the publication of several edited collections (Lawrence *et al.* 1996; Lockie and Bourke 2001; Pritchard and McManus 2000) analysing the impacts of changes that appeared to many, especially to those living through this change, to be unremitting. Research examined the effects of globalisation and economic restructuring, demographic and socio-cultural changes, concerns with government and environment. Running through many of these accounts was a theme of 'loss': loss of industry, businesses, services and people. Encompassing much of this was the loss of a rural destiny that was assumed would be available for future generations but, as the 1990s progressed, became alarmingly ephemeral for many individuals, families and communities. That extended into a general feeling of abandonment by government to the vagaries of 'markets', culminating in the One Nation 'backlash' in 1998 (Pritchard and McManus 2000), a conservative populist and protectionist response to the perceived neglect of 'ordinary Australians'.

The theories that informed research on rural Australia during this period run the full gamut of influences, ranging from the political-economic work produced by the likes of Gray (1991), Lawrence (1987), Pritchard (1999) and Tonts (2001), to the feminist perspectives offered up by Alston (1995), Liepins (1998, 2000) and Pini (1998), alongside a cultural turn that encompassed a broad range of issues from the cultural industries (Gibson and Davidson 2004) to the culture-nature nexus (McManus 2008). However Australian rural studies have not only drawn on the work of others when applying this 'diverse and vibrant blend of theoretical … perspectives' (Tonts *et al.* 2012: 291), they have also been at the vanguard of important theoretical developments. For instance, Holmes' (2006) development of the multifunctional rural transition responded to the conceptual simplicity of the post-productivist countryside and rethought the approach so that it better accounted for the diversity and complexity that characterised the Australian rural experience. Similarly, the work of Argent, Cheshire, Higgins and Lawrence individually and collectively (Argent 2011, Herbert-Cheshire 2000, Cheshire 2006, Higgins 2001, Herbert-Cheshire and Lawrence 2002, Herbert-Cheshire and Higgins 2004, Cheshire *et al.* 2007) pioneered explanations around rural governance using the Foucaultian 'governmentality' framework. These evolving theoretical influences permeate the work of the chapters that follow.

More than a decade into the twenty-first century and the theme of 'change' is no less significant when considering rural Australia. However, this most recent period of change in rural Australia is distinctive in important ways. Since the turn of the twenty-first century new and significant demographic, economic and environmental changes and challenges facing rural Australia have emerged. These issues range from the challenge of ageing populations, rural youth outmigration, changing immigration policies (that seek to better distribute the abilities of skilled-immigrants into rural Australia) and new governmental relationships with Indigenous populations. The restructuring of agricultural industries has continued, however the economic story in rural Australia (and Australia generally) is now centred around what has become known as the 'patchwork economy' or the 'two (or even three) speed economy'. This seeks to encompass the unimaginable wealth that the mining boom potentially offers some regional areas, while at the same time recognises the on-going issues of regional economic decline in others. In some places this has meant the emergence of a 'multifunctional economy', as services take over from production. As an economic space, rural Australia is increasingly not only both a place of production of agriculture and minerals but also an idea that individuals seek and are encouraged to consume. Environmentally rural Australia has been faced with the socio-economic implications of drought, water rights, mining and changing farming practices, which have prefaced new social, cultural and economic reforms. Indeed, the rapid pace at which these changes have occurred – notably with the expansion of mining – means that environmental issues have not only achieved new significance but there has been concerted regional resistance to land sales and coal seam gas (CSG) mining. Economic growth and decline in rural Australia, as in many parts of the world, have become issues of national importance (Cocklin and Alston, 2003). Despite the increasingly coastal location of the nation's population, issues of rural change in Australia continue to be debated, discussed and viewed as critical in terms of national development.

A second theme also emerges from the patchwork economy and into this text: diversity. Rural and regional Australia is a unifying phrase for an extremely variable collection of people, economic activities and environments – a unity sometimes only thinly sewn together. More than in most developed countries Australia is 'a land where regions matter' (Beer *et al.* 2003: 1) and where tyrannies of distance remain very real. Real variations exist within Australia over their importance. Queensland, for example, with several large cities, is more regionally-oriented than New South Wales, where NSW is sometimes seen as merely an acronym for the largest coastal cities of 'Newcastle-Sydney-Wollongong', separated by a 'sandstone curtain' from the bush. In Australia, regions have often become identified with the emerging concept of a somewhat disadvantaged 'rural and regional Australia' (Pritchard and McManus 2000, McManus 2005). The existence of significant, and growing, variations in regional prosperity has long prompted calls for Federal Government involvement in regional development, on the basis that it has the financial capacity, due to its revenue generation, to overcome state and regional conflicts (Maude 2003, Beer 2006, Brett 2007). The Federal Government, under the Liberal Party Prime Minister, John Howard, which governed for eleven years until 2007, in coalition with the rural National Party, shied away from involvement. Ironically, its successor, the Australian Labor Party government, has had to take a greater interest.

Restructuring of traditional agricultural industries and a resurgence of mining have also created 'winners' and 'losers' – further fragments in the patchwork. Certain regions, such as parts of Queensland and Western Australia, where mining has boomed, and Victoria, with better soils, more space, and high capital-intensive forms of production, have improved their economies of scale, gained export markets and retained or even boosted their populations. In other regions where farms and exports are smaller, and where capital is scarce, local economies have stagnated. Patterns of growth and decline and resultant social problems have not been distributed evenly between, or within, regions. While some centres continue to grow, many smaller towns have become caught in a vicious cycle of decline, losing residents, industries and confidence about prospects for a sustainable future and fearing neglect from the centre. Above all rural and regional Australia has become substantially more complex than in any other era.

The collection of chapters within this volume explores the above themes of change and diversity through the three intersecting areas of population, economy and environment.

Peopling the Australian Countryside: Is there a Next Instalment?

How the national population is distributed has been a long-term and ongoing concern in Australia. Complex demographic changes are producing uneven and challenging circumstances for rural and regional Australia. Counterurbanisation processes that began in the 1970s and a long-term trend of rural youth outmigration are perhaps the most familiar of these changes (Hugo and Smailes 1985, Argent and Walmsley 2008, Chapter 2). Rural and regional locations experience these changes differently. In particular there is a growing division between coastal and peri-urban areas in eastern and south-western Australia and inland, predominantly farming, regions. The former tend to struggle to meet the needs of burgeoning populations, while the latter areas contend with the consequences of population decline. Regions within a hundred kilometres of the coast, such as Margaret River (WA) and Kangaroo Valley (NSW), have thrived. Distant and smaller places have struggled. Despite many subtle changes population growth in Australia has remained close to the coast, and intensified environmental pressures in many places, in terms of pollution, reduced habitat and biodiversity, pollution of waterways and simple loss of open space (Danaher 2008) in areas that are already subject to pressures from coastal erosion. In the coastal regions of Australia urbanism and urbanisation are creeping into spaces traditionally classified as rural and regional. Elsewhere, regardless of population size and trends, rural and regional areas in Australia are coping with the population challenges of ageing populations, declining fertility, increasing demand for younger/working-aged members of the population, labour shortages and the need for improved service provision, in a thinly populated continental inland.

Young women have been an especially mobile group, resulting in the popularity of reality television programmes such as *The Farmer Wants A Wife*. In Chapter 2 Dufty-Jones, Argent, Rolley and Walmsley examine the gendered dimensions of the 'cultures of migration' that inform decisions to leave and/or return to rural locations, particularly those cultures that influence young rural women. They find that female role models, both positive and negative, are critical in influencing female out-migrants' decisions to leave. Similarly, gender is found to be an important dimension influencing the return migration aspirations of former rural out-migrants. Young rural women identified more barriers to making a successful return move compared to their male counterparts. Dufty-Jones *et al.* point out that the gendered differences in the cultures of out- and return-migration are significant, with rural communities potentially missing out on a considerable section of the human capital that can be generated through return-migration.

In large parts of regional Australia traditional agricultural life has continued to contract – with smaller populations in declining towns (Forth and Howell 2002). Rural populations are also ageing, both through selective migration and greater longevity. This has created additional pressures on many rural centres where services are declining yet the needs of an ageing population of greater longevity are growing. Sustaining a minimum level of services and facilities in declining towns in an era of privatisation is challenging. In various parts of the country the infrastructure and community life of many rural and remote towns has slowly disappeared through bank branch closures, the loss of football and cricket teams, and declining populations in some areas (Argent and Rolley 2000). Once towns can no longer field sporting teams not only are there fewer activities for youth, but ancillary social activities also disappear (Tonts and Atherley 2005, Atherley 2006, Spaaij 2009). Sport is not only a visible symbol of rural wellbeing but it plays an integral role in the development of social capital. Volunteerism in sport, and other social organisations, such as choirs and churches, is a vital part of the functioning and character of a community. Disappearing sports teams, and amalgamated or empty churches, demoralise communities more than the loss of banks. Those who remained increasingly railed against boredom.

As some rural populations have declined various attempts have been made to 're-balance' the population. Yet only one inland city has grown rapidly: fifty years after it was established as the national capital, Canberra became the largest of Australia's inland cities, its artificiality indicative of the problems of regenerating regional Australia. Other more subtle strategies, from providing rural relocation grants, renting empty farmhouses or developing 'Evocities' have all had their moments (Connell and McManus 2011), but it is readily apparent both that counterurbanisation in most forms is directed to the coasts and that the resultant sea change absolutely dominates inland tree change. Only 'micropolitan' centres – the sponge cities – have thrived inland, yet keeping small towns alive is vital to securing a rural future (McManus *et al.* 2012, Martin and Budge 2012). The chapters by Drozdzewski (Chapter 5) and Davies (Chapter 3) develop these different dimensions of rural population change in Australia. Presenting the tree-change experiences of two very different regions in rural NSW, Drozdzewski demonstrates the diversity of reasons why individuals pursue a tree change, where such moves emanate and the impacts of this population change on local communities. A hybridised vision of rural-living emerges from the accounts of both in-migrants and long-term residents, while the aspiration for economic development and services growth is delicately balanced against the equally significant asset of an idyllic rural lifestyle and a landscape that provides an unchanging link to an agrarian past.

Davies explores tree change migration through the important demographic trend of an ageing population. She examines how elderly urban to rural migration flows are influenced by the social, economic and geographical factors of the destination community. Elderly migration flows build over time and influence the socio-economic functions and fortunes of rural towns as in the case of Gloucester (NSW) where an ageing population provokes important structural economic and social change. Further economic and social change should be anticipated and can be capitalised on by the local community.

Juxtaposed against such internal migration patterns and their consequences is a 'paradigmatic shift' in immigration to Australia, with an increasing proportion of immigrants from Asia, more flexible/temporary migration and a greater likelihood of new immigrants settling in non-metropolitan locations (Hugo 2004, Wulff *et al.* 2008). Hugo charts what this change means for Australia's rural regions in Chapter 4. Reflecting on the historical role of immigration in population change in rural and regional Australia, Hugo shows how the 'regionalisation' of immigration policy has become increasingly linked to regional development in rural regions. The chapter contributes to an emerging literature on ethnic diversity in rural Australia (Jordan *et al.* 2009; Panelli *et al.* 2009; Dufty and Liu 2011), even though, the recent phenomenon of increasing numbers of immigrants settling in rural regions 'has largely gone under the radar of researchers and policymakers' (Hugo and Morén-Alegret 2008:474).

Whereas gloom and doom hang over the demography of regional white Australia, despite attractive and welcoming images of a bucolic countryside (Chapter 11), where populations are ageing and moving away, the converse holds for indigenous Australia. In Chapter 6, Taylor shows that for many Indigenous populations there has been a continued shift to the bush, even towards some of the remotest places in Australia, while fertility rates (but also mortality rates) remain high. While urbanisation of the indigenous population has continued, but into relatively small places like Thursday Island and Wadeye, where Indigenous people predominate, there has been a degree of rejection of the assimilationist doctrines of the past, and a strengthening of homelands. However welfare dependency is real, and interventions have sought to improve access to (and ownership of) formal housing, employment, education and health care. Deeply embedded moral contests about values and ways of living are daily being played out in remote Australia. Developing infrastructure and services in sparsely settled Australia has necessarily been especially difficult and costly, partly because of the rising costs of service provision in an age of public austerity, but especially because of the particular challenge in ensuring adequate human resources for education, health and other services. Numerous special schemes have been devised to recruit, train and allocate workers from, in and to rural areas, but with only limited success Government interventions have rarely been sustained and have too often changed purpose and direction, as regional Indigenous Australia remains a zone of post-colonial adjustment, where the order of kinship and the order of the market exhibit a long, slow friction (Austin-Broos 2009). In a much deeper sense it constitutes a powerful reminder of the diversity of regional Australia, not merely in its demography.

It's the Economy, Gina!

If Australia has become a nation of patchwork economies – two and three speed regions – defining and differentiating these distinctions is difficult. Crudely the economy is booming in the north and west with a vibrant mining sector and experiencing relative stagnation in the south and east – where manufacturing industry has struggled – but with enormous subtleties. Western Australia (WA) above all has achieved rapid economic growth on the back of the mining industry – centred in the Pilbara – that has drawn in several thousand international migrant workers. The mining boom took the dollar to unprecedented heights, pushed up the exchange rate and contributed to a two-speed economy that has reduced some rural exports and inbound tourism. Agriculture and mining have experienced waves of foreign investment, and new export booms, both dominated by Asia. China alone went from taking 11.5 percent of Australian exports valued at $16 billion in 2005 to 27 percent in 2012 valued at $73 billion. Collectively China, Japan, Indonesia and South Korea take 44 percent of exports, ranging from iron ore and gas to beef and cheese. Scarcely surprisingly Prime Ministers make regular reference to the Asian Century.

The gravitational attraction of the north has shifted the demographic balance from South Australia and Tasmania towards Queensland and Western Australia. Between 2001-2011 Queensland had the greatest growth (up by 845,200 people), while Western Australia had the fastest growth, increasing by 24 percent, followed by Queensland (23 percent) and the Northern Territory (17 percent). Victoria and the Australian Capital Territory grew by 15 percent, the Australian average. The remaining states all had growth below the Australian average, with New South Wales at 10 percent and South Australia and Tasmania both at 8 percent. The mining boom emphasised the differences.

Mining

Nothing has changed the economic geography of twenty first century Australia more than the resurgence of mining, especially in Western Australia and

Queensland. Mining of iron, coal and other minerals, mainly destined for Asian markets, but especially China, has generated enormous incomes (making Gina Rinehart, the owner of Hancock Prospecting, reputedly the wealthiest woman in the world) and resulted in ongoing debates over the taxation (and ownership) of mining companies, employment generation (and whether that should or could be Australian, including local indigenous people, or from overseas), the appropriateness of uranium mining, infrastructure provisions, and mining's pressure on the environment, notably in demands on water and pollution of water sources, and on ancient Aboriginal paintings and sacred landscapes (in parts of remote WA and Queensland). It has even resulted in debates over state rights, partly centred on who actually owned the nation's mineral resources. Such debates were particularly vigorous in WA, the main locus of the mining boom, when (until late in 2010) WA was the sole non-Labor controlled state in a nation that had a Labor government. Indeed WA had a history of difference from the east, of antagonism, of detachment (prior to the mining boom) and even of secessionism. Conflicts over land use, water and the role of mining were most important in more populated regional areas, such as the Hunter Valley, where other valuable economic activities and greater population concentration existed, and that came to a head with the massive upsurge of prospecting for coal seam gas (CSG) after 2010. In Chapter 13 McManus examines this conflict over land, water, access and the identity of the Hunter region. These conflicts have intensified with the introduction of CSG as another land use in an environment already affected by extensive coal mining operations. McManus reflects on how both the internal diversity of the Hunter region combined with the impact of transboundary flows (occurring at a variety of scales) is for the most part overlooked in discussions around the 'patchwork economy'. He argues that this has significant implications for the social, economic and environmental sustainability of the region.

Developing a labour force for mining in remote areas has proved difficult. For instance, the Pilbara, in the north of Western Australia, has become the new frontier in the resources boom (Plummer and Tonts 2013). However the region lies in one of the hottest parts of inland Australia, services are poorly developed, and workers are unwilling to move into non-mining jobs at standard wage rates. Difficulties in attracting labour to this region are compounded by the fact that Western Australia has long experienced almost full employment. In 2012 the WA unemployment rate was 3.8 percent, its lowest for four years, and well below the national unemployment rate of 4.9 percent, itself the envy of most other nations. While Australian manufacturing has experienced a slow decline, such percentages indicate only a small number of skilled workers available to work in the resources sector, and a well-established national skills shortage. Not surprisingly, by 2012, a substantial proportion of the holders of 457 visas (issued to overseas skilled workers) were in the resource sector. Efforts to lure unemployed manufacturing workers across the Nullarbor have largely been unsuccessful, despite high wages and escalating demand. The reluctance of workers to travel from the south-eastern states for work in the north, has even resulted in FIFO workers in WA, a self-defined labour aristocracy, even described as 'job snobs' and 'princesses'. Moreover the West Australian Premier, Colin Barnett, has said:

> The fly-in fly-out workers are the modern heroes. They do separate from their families, they do put up with some loss of amenity, and they go up there and they work in harsh conditions for long hours. They do exciting work and they are building this state and building the nation (quoted in *The Australian*, 7 March 2013).

Meeting the needs of remote mines in both WA and Queensland has meant the substantial growth of Fly-In-Fly-Out (FIFO) and Drive-In-Drive-Out (DIDO) schemes, with a workforce sourced hundreds of kilometres away. In WA for example workers travel from at least as far as Bali – a unique Australian example of international commuting. This has boosted some small private airlines, but with damaging environmental and especially social consequences. Despite labour shortages in the new mining areas, rapid growth has attracted enough key workers from elsewhere in the mining states, resulting in considerable ire in towns struggling to hold on to skilled workers (Connell and McManus 2011). Mining created labour vacuums in areas such as the Atherton Tablelands of north Queensland, where the sugar industry was bereft of labour, as nearby mines at Chillagoe and FIFO mining from Cairns, drew workers attracted by much larger wages. Indeed in Queensland mines are regarded as the biggest culprit in 'poaching' workers (Chapter 7) while the sugar industry has turned to Pacific islands migrants as it had done a century earlier (Chapter 4).

Conversely FIFO into established mining towns like Mount Isa and Broken Hill has discouraged permanent populations and reduced social life to the extent that the mayor of Kalgoorlie has called FIFO/DIDO 'the cancer of the bush'. The mining towns of Kalgoorlie, Broken Hill and Mount Isa have all lost population in this century; the 'Iron Triangle' towns of Whyalla, Port Augusta and Port Pirie, lost 10,000 people in the space of a generation, after 1986. Not only have these towns become less able to serve the needs of a wider area but the innovations associated with new mining technology provide no benefits to their developing a post-mining economy (Martinez-Fernandez *et al.* 2012). Mining booms generate the classic resources curse ('Dutch disease') problems. Any real sense of sustainable mining, or broader social and economic development, has flown away. McIntosh and Carrington emphasise these social challenges in Chapter 7. Examining mining industry workforce practices in a frontline rural town in the Bowen Basin of Central Queensland, they found the presence of large numbers of FIFO and DIDO workers had dramatically transformed the community. While some sectors of the local economy (e.g. hotels, service stations etc) benefited from the expansion of mining, various negative impacts included: the loss of individuals that contribute to sporting clubs and community volunteering activities; pressure on local infrastructure and services (e.g. housing, hospitals etc.); and the impact of alcohol- and drug-fuelled violence. While the boom in mining should continue, little has been done to plan for how frontline rural communities can reap benefits from it.

Agricultural Restructuring

Drought and commodity prices have conspired to challenge agricultural development in settled Australia. This book was written in the wake of the longest drought to grip Australia for the best part of a century. Yet, at the same time, it is certainly not environmental factors that have been the sole influence on agricultural change; 'it's not just the drought' (Chapter 12). Fluctuating international export markets and global prices are critical determinants of all the key exports such as wheat, dairy products, beef, lamb and wool. Prices hinge on Asian economic growth, as Asian markets have gradually replaced more distant European markets. Demand for beef in Asia is partly fuelled by the fast-food revolution and especially the rise of McDonald's. That in turn has influenced Chinese (and other) interests in purchasing Australian farmland. Despite growing Asian demand for 'traditional' products, agriculture has constantly changed, steadily evolving away from the broad-acre farms of last century – where wheat fields and 'living off the sheep's back' were still the norm – towards more complex agricultural systems as family farms have slowly declined but niche agricultural products become more important. Sheep numbers in NSW have declined to their lowest levels in a century, whereas in Victoria sheep production has thrived but citrus production slumped. In NSW cattle production has fallen and many abattoirs closed in the last two decades, although the area of wheat cultivation has expanded. Even within states and in one sector there are complex and ever-changing patchworks.

Agriculture has come under pressure in most parts of the country, from urban expansion, price fluctuations (and trade negotiations), environmental concerns, and even the manner of slaughter in Indonesian abattoirs (echoes of sheep transport to the Gulf). The banning of live cattle exports to Indonesia in 2012 temporarily paralysed the export industry. Farmers are also ageing; the average age of Australian farmers in 2010 was 56 and their children are not necessarily taking over. As McKenzie shows in Chapter 8, in an era when technological expertise is becoming crucial, farm succession is often in doubt. The future of family farms has become critical as children are increasingly unwilling to take over from parents; such situations have discouraged investment in technology, stagnation if no successor is apparent and land sales (Wheeler *et al.* 2012). McKenzie notes that the need to do more with less has in part contributed to the decline of the mixed farm. As part of this process of increased specialisation, farms have gradually been consolidated into fewer and larger units marked also by a shifting of economic functions from smaller to larger settlements (Pritchard *et al.* 2010). However the shift away from the mixed farm comes at a cost in terms of the financial, social and environmental sustainability of the farm as the resilience of the enterprise in all these respects is eroded through specialisation.

Salinity has affected some inland areas. The probability of adequate future water supplies has challenged others, and resulted in a shift away from horticultural production (e.g. grapes, stone fruit, citrus etc), where water could no longer be guaranteed (Chapter 9). This kind of environmental uncertainty is greatest where agriculture is dependent on irrigation, especially in the massive Murray-Darling basin where much of the present century seems to have occurred in the shadow of both drought and the necessity to develop more effective means of water regulation, that balance economic, social and environmental objectives (Chapter 12). Throughout Australia the kind of cumulative uncertainty that has come from such multiple, variable, and often unpredictable, influences has raised issues of farm succession and deterred investment, as the viability of agricultural futures can seem threatened. This has threatened the vitality of many small farms and the mantra 'get big or get out' has defined recent years (Chapters 8 and 12). Agriculture has become more specialised, except in boutique metropolitan penumbra locations, as production has demanded superior technology, knowledge and skills.

Where agriculture retained vitality, production often changed. Boutique wineries increased in numbers and wine production soared, at least until the global financial crisis and gluts on overseas markets. Wine exports have subsequently gone into freefall, hit by a 'perfect storm' of a strong Australian dollar, global recession and growing overseas competition. There has been an increasing agricultural diversity, ranging from olives, gourmet food (including cheeses and organic products), and cool climate wines in upland areas, such as Orange and Stanthorpe. Organic agriculture has boomed with farmers markets springing up in most urban centres, and food festivals scattered over the spring, and sometimes autumn, countryside. A preference for 'local food' whether because of notions of taste and freshness, or environmental concerns over food miles, has transformed parts of rural Australia within easy reach of metropolitan capitals. Several such areas have simultaneously become tourism destinations, marked by the rise of wine tourism, as in the Barossa, Hunter Valley and Margaret River regions, and become the favourites of certain magazines and restaurant menus.

As the orientation of agriculture to Asia increased, so Asian investment in agricultural land also increased; land tenure became a significant topic of debate with national media querying who 'owned the farm' and many fearful of a 'Chinese land grab'. The Federal Labor government was even accused by the National Party of 'provocatively and dangerously' favouring Chinese bids over local contenders in the race to buy and develop large tracts of farmland in the Ord River in northern WA to grow sugar for ethanol. The National Party Senate leader, Barnaby Joyce, provocatively observed: 'I don't blame the Chinese for being prudent and planning ahead for their own food security, but if this is in their best interests to buy our farms than it can't be in ours' (quoted in *The Australian*, 1 June 2012). A register of foreign land ownership was considered. Just 0.1 percent of total direct foreign investment was then in agriculture, forestry and fishing, while 89 percent of agricultural land was entirely Australian-owned. Controversy accompanied the 2012 sale of Cubbie Downs station in southern Queensland, the largest privately owned irrigation property in the southern hemisphere, after it had gone into administration towards the end of the drought years. It was sold to Shandong RuYi, a predominantly Chinese clothing and textile company. The sale of such a large property raised concerns over foreign ownership and further questions about the use of Murray-Darling water for cotton irrigation in Queensland at the expense of downstream users.

Like mining, agriculture has also pushed northwards, both parts of the long-held and regularly frustrated quest to 'open up the north', symbolised by the opening of the Darwin-Adelaide railway line in 2000. The Ord River scheme, begun in 1959, appeared closer to some success after half a century. Like many other experiments in tropical agriculture it initially failed because of difficulties growing crops, attacks from pests, salinity and erosion, but fruit, vegetables and even sandalwood were exported to South East Asia. In 2012 the Chinese property development conglomerate, Shanghai Zhongfu, purchased the sole right to develop 15,200 hectares of land, to construct a sugar mill, grow four million tonnes of cane a year and export 500,000 tonnes of sugar. Previous efforts to produce sugar cane in the Ord River region failed because the scale of operation was too small, given the distances and costs involved. The Chinese investment caused controversy on two fronts, confronting the old idea of the Kimberley region as Australia's 'food bowl' to produce substantial crops, where state and federal governments had developed the infrastructure, and selling more Australian agricultural land to foreign investors. However this was part of a broader push to increase Australia's role in supplying global food markets, symbolised in the joint Australia-China report (Australian Government 2012), that emphasised the value of Chinese investment in Australian agriculture. Slowly the Ord area was taking the lead in becoming a small part of a possible Asian 'food bowl'.

Technological investments have reduced demands on labour, although the ageing of farmers has meant new managerial demands. Shortages of agricultural workers are widespread, which has led to some changes in gender relations as women take over erstwhile male activities such as shearing. Even so, in many areas, especially in horticulture and fruit farming, agricultural workers are scarce. Fruit picking areas have long been dependent on backpacker labour, but continued shortages and concern over the reliability of backpackers, resulted in Australia establishing a Pacific Seasonal Worker Pilot Scheme (PSWPS) in 2008, bringing workers from Pacific Island states on short six-month contracts (Chapter 4). Most worked in Victoria, NSW and increasingly Queensland, where farms had lost workers to mining. As migrant Melanesian workers even returned to the Queensland cane fields there were uneasy resonances with nineteenth century labour migration, especially from Vanuatu (then the New Hebrides) to work on the sugar plantations: a process and era known as 'blackbirding' (Connell 2010).

In the 12 years between 1997 and 2009 Australia's agricultural land area declined by 11 percent from 462 million hectares to 409 million hectares (Budge *et al.* 2012). Only about 3 percent of Australia is actually suitable for cropping, and even less is considered prime agricultural land. Some of the best agricultural land is being lost to urban sprawl, to mining, and even to national parks and conservation reserves, with accelerating pressure in such areas as the Yarra and Hunter Valleys. Since Australia has the sixth largest land area and the lowest population density of most developed nations, and an economy that is not wholly dependent on agriculture, the question of whether or not there will be sufficient good quality land available for agriculture in the future has never been a high priority issue. Current disputes about future land use are thus concentrated in populated areas not far from metropolitan centres. Low density urbanisation is resented. Pressure on agricultural land for mining and other purposes has resulted in strong reactions, and particularly strong opposition to CSG mining both on the fringes of Sydney and in agricultural regions like the Hunter and Manning River valleys and plains. A new militancy has stirred in conservative rural areas.

The Lock the Gate Alliance began in 2010, across NSW and Queensland, coordinating local groups, and opposing CSG mining and other 'rapacious mining industries' on environmental grounds (the potential pollution of rivers and aquifers) and the loss of valuable farmland to mining. Its website declares that its mission 'is to protect Australia's natural, environmental, cultural and agricultural resources from inappropriate mining and to educate and empower all Australians to demand sustainable solutions to food and energy production. We are working in our local communities: home by home, road by road, to lock the gate and protect our land, water and future'. Unlike most mining booms, and especially that in WA and Queensland, the potential CSG boom involves land that has multiple other uses and where many people live. Mining now raises more complex challenges on a much wider scale than in the Hunter Valley, where tensions and disputes have long been present. In this century social, economic and environmental issues – and therefore political debate – have all become more diverse and more complex.

'Of Droughts and Flooding Rains'

While drought accompanied the first decade of this century in much of regional Australia, the second decade, thus far, has been accompanied by floods in several states (notably Queensland and Victoria) that have devastated some places, and raised new questions about the significance of extreme weather conditions, often considered to accompany early phases of climate change. At the beginning of 2013 rural and regional Australia witnessed some of the hottest temperatures on record (49.6 °C in Moonba, South Australia), extensive bushfires around Coonabarabran (NSW) that destroyed over fifty properties, record-breaking floods in Bundaberg (Queensland), and the disconcerting non-appearance of the 'wet season' in Darwin (Northern Territory). In low-lying areas, notably in the Torres Strait, sustained flooding heightened concerns over climate change (Green *et al.* 2010). The environment has always mattered in rural Australia but today it is as important – and perhaps uncertain – as it has ever been.

Drought brought complex human consequences, far beyond the collapse of agricultural production, and continued uncertainty. The mental health of rural residents worsened and suicide risk increased following heightened uncertainty and stress in regions where suicide rates were already above national norms (Hart *et al.* 2011, Guiney 2012). Certainly the climate has played an unusual role in regional development in recent years, at a time when environmental factors are generally becoming of greater importance in conceptualising rural development, whether to strengthen the environmental component of the triple bottom line or to develop a stronger sense and practice of conservation agriculture. Climatic hazards have contributed to the uncertainty of regional development and climate change may pose new threats in the not too distant future.

The most recent experiences of drought across Australia has brought about a paradigmatic shift in the way in which rural sustainability is understood. The chapters by Askew, Sherval and McGuirk (Chapter 12) and Smith and Pritchard (Chapter 9) examine this change from two different angles. Askew *et al.* identify in Chapter 12 how policy approaches to droughts in Australia have shifted away from viewing these events as unexpected and intermittent to instead being understood as natural and routine. The way drought is understood, experienced and managed today means that it is an important (but far from independent) element in the environmental, economic and socio-demographic changes that are fundamentally and permanently changing the farming practices and structures in Australia.

The experience of severe drought has only served to underline the importance of access to water This century has witnessed seemingly constant debate in one place or another over the future of rainfall, climate change, irrigation, river systems and dams, even amongst urban Australians who had no previous interest in such matters. It has also marked the entry of the Federal government not only into debates about the management of water systems but towards the greater national control and regulation of water. Water was now too important, and too many people and activities were involved in its use and regulation, for this to be solely a state and local affair (Chapter 9). The economy (and a steadily growing national population) depended more on water than at almost any other time in the past.

Throughout this century complex debates have centred on the supply, allocation and use of water – especially as the drought took hold – prompting arguments about land use (were irrigated cotton and rice efficient uses for scarce water?), how might water from the major Murray-Darling catchment be allocated equitably, to farms, residents and environmental ends, when it involved four quite different states – and the national government – and when in some years the river barely reached its mouth. Attempts to develop an acceptable plan in an increasingly user-pays context could never satisfy a continuum of interested parties, from irrigators who sought more water to environmentalists who sought to retain more water in river systems and wetlands (Chapters 9 and 12). In the Murray-Darling basin:

> Irrigators upriver are at loggerheads with others down river; the battle between graziers and irrigators has spilled into violence; some environmentalists won't be satisfied until river farming goes away altogether; environmentalists and graziers have formed an unusual alliance against cotton farmers ... And the irrigation lobby warns that shrinking their overall entitlement risks not just their livelihood and national output but also the wellbeing of hinterland communities, which would shrivel as jobs and cash evaporate (Humphries 2012: 6)

In almost every part of Australia, and at variable scales, debate ensued about 'drought-proofing' Australia and how that might be accomplished – by dams or by demand side constraints, that often brought regional water authorities into conflict with local residents and scientific observers, as in debates over a possible Tillegra dam in the Hunter River valley (Sherval and Greenwood 2012), which emphasised the diverse range of planning challenges in seeking to achieve a balance between managing rivers as ecosystems, water supplies (for agriculture, mining and residents- near and far) and as an environmental asset, providing largely unmeasured and immeasurable environmental services. 'Enviro-mental-ists' are deeply resented and distrusted in irrigation areas. Achieving agreement over the right balance of land and water uses has been as elusive as the rains, and disputes have occurred in every place and at every scale. When a plan for the Murray-Darling finally emerged in 2012 (Chapter 9) it was no surprise that in downstream Victoria it was seen as a 'death warrant' for agriculture. Smith and Pritchard (Chapter 9) trace the way in which water and its relationship to rural Australia has been fundamentally rethought through the way the Commonwealth Government, as a market participant, has legislated to re-distribute water from agriculturalists to the environment. This has also been accompanied by substantial government support for farmers to find new ways to manage water. Both approaches however challenged the long assumed priority over most of the twentieth century that agriculture had access to water resources at the expense of all else. Exacerbated by drought and climate change, the environment is integral to the ongoing structural reform of rural Australia in the twenty-first century.

Changing resource uses – and the rise of mining – affected the environment in complex and unanticipated ways. The expansion of mining in Queensland meant more demands for infrastructure, larger ports and more movements of larger ships – putting intense pressure on the coast to the extent that UNESCO regarded the Great Barrier Reef as moving towards being 'in danger' – threatened by pollution and siltation, again testing state-federal relationships. In large areas of NSW and Queensland communities rose up against the threat of CSG (though many others were willing to sell or lease their land for profits that went far beyond what agriculture appeared capable of). A fluctuating and perhaps permanently changing environment affected more regional lives while economic change brought new threats to the environment.

A New Countryside?

Despite multiple changes in rural and regional economies the visible landscape, other than in mining areas and on metropolitan fringes, has scarcely changed. Though, where windfarms have been built, there are hints of alternative diversities. Yet, ubiquitously, regional Australia has become a multifunctional rural area – a place of production, consumption and protection/preservation (Holmes 2002, 2012). Rural Australia is presented to and consumed by Australians in a variety of forms; from outback isolation to the gentility of a English rural idyll it appears that rural Australia can cater to your taste. One particular 'vision splendid' of rural Australia is presented in Chapter 11, where Connell examines the particularly urbanised vision of rural Australia, through a discourse analysis of the magazine *Country Style*. As Connell notes, it is much easier to find *Country Style* in middle-class suburbs of Sydney than in the newsagents of small country towns. *Country Style* simply provides an insight into how one segment of Australian society would like to imagine, experience, reproduce and consume rural Australia, though it represents an elite vision of rural Australia disconcertingly disconnected from the real lives of most rural Australians.

Rural and regional Australia have entered a post-productivist era where agricultural production is no longer the sole economic activity. Diversity has been crucial especially during the drought years. Domestic tourism, however, once a standby of rural Australia has stagnated. Half a century ago coastal Australians often holidayed inland, and farm tourism flourished. That era has gone. Australians stick to the coast or go outwards, for cheap holidays (e.g. Bali, Fiji) rather than inwards. Cheap fares and a strong dollar have accentuated that trend. Conversely the strong dollar has discouraged international tourists, disadvantaging regional centres dependent on tourism, notably in north Queensland and in the 'red centre'. While tourism from China has remained strong, this has been almost entirely metropolitan based, and has done little for growth in regional Australia. Here regional Australia would again like to turn to Asia but success is far from guaranteed.

While tourism may have slumped, a new rural creativity has drawn visitors, on a relatively short-term basis. As Gibson outlines in Chapter 10, craft markets and festivals, even the most improbable, have emerged and proliferated, stimulating tourism and regional development and drawing rural communities together. In Australia, more than in other countries, festivals are antidotes to drought, depression and part of the regeneration of the countryside. The Gympie Country Music Muster, reasonably typically, has brought new incomes to local communities, stimulated a stronger sense of local identity and established rural places on urban mental maps (Edwards 2012, Gibson and Connell 2012). Yet, as Gibson concludes, in the end no one festival, or even clusters of festivals of different kinds (and most small towns have several), can be a panacea for rural decay. Some festivals fail. Yet again there is an inherent geography, where the most successful festivals are within a few hundred kilometres of big cities (and the coast), as new mobility has limits. While festivals enable a limited renaissance there are limits to festival growth, because of competition, accommodation constraints, in some cases too frequent repetition and high prices, and, for some classical festivals especially, the difficulty of finding appropriate venues.

Small towns are thoroughly capable of producing creativity, despite assumptions that creativity – and the creative industries – are metropolitan phenomena, and that 'best practices' are to be found in the largest cities, from where they may trickle down (Chapter 10). Festivals alone demonstrate that this is no longer tenable, and that many small towns have been able to gain significant economic and social benefits by developing and trading on improbable, even wholly fictitious and sometimes 'unworthy' events and associations: strategic inauthenticity. Festivals, diverse and creative in themselves, are only one part of a wider creative scene, that embraces the Deniliquin Ute Muster, a 'kombi' muster at Old Bar (NSW), goat races at Barcaldine (Queensland) and the 'running of the sheep' at Boorowa (NSW). 'Utes in the Paddock' (NSW), where a '7m metal marsupial is spectacularly silhouetted against the burnt blue outback sky' (Davies 2009), the Tin Horse Highway (WA) and Silverton (SA)'s Mad Max Museum are further tiny components of the quirky and original rural creativity that has gradually transformed image and reality in regional Australia. While the countryside may look the same it functions in new and necessary ways.

Conclusion

'Rural and Regional Australia' is a slippery geographical category, coming to encompass all of non-metropolitan Australia (each state having only one major city), within which there is great diversity: broadacre farming regions involving the production of cash crops at scales of thousands of square kilometres; regions producing rice and cotton with state-sponsored irrigation; coastal agricultural zones with smaller and usually older land holdings (decreasingly the places of traditional 'family farming' communities); single industry regions focused around minerals extraction or defence (many of Australia's major defence bases being located outside state capitals either in sparsely populated regions in Australia's north or in smaller 'country towns' in the south, where they dominate local demography); semi-arid rangelands regions dominated by enormous pastoral stations leased on Crown land (single examples of which rival the United Kingdom in size); and remote savannah and desert regions many thousands of kilometres from capital cities, supporting Aboriginal communities living on traditional country mixing subsistence hunting and gathering with government-supported employment and food programmes.

All of that has been complicated in a new century of remarkable economic globalisation – trade, investment, tourism, foreign ownership in mining and agriculture and growing pressures for both migration to Australia and from within Australia for skilled workers. Agriculture has given way to mining as the basis of regional growth, but both are increasingly linked to the dynamism of Asian economies – the main export destinations – in the 'Asian Century'. Part of the future of agriculture depends on Asian investment, in established agricultural areas and in the 'new' agricultural regions of the deep north. For a rich-world country Australia remains unusually dependent on the exports of commodities – minerals and agriculture – and thus on the vagaries of demand, foreign exchange rates, political whim, indigenous rights and tenure, and other factors beyond national control.

For all that some magazines portray images of a never-changing Australia where heritage dominates, rural and regional Australia is ever changing, but with extraordinary diversity: multiple parts and multiple directions, patchworks and variable speeds, intersecting places of population decline and growth, mining, agriculture and constantly evolving service economies, with fading public sectors – hence heated debates over service and infrastructure provision – and occasionally dynamic private sectors. In certain regions, notably those not far from metropolitan Australia, such as the Hunter Valley, multiple sectors (tourism, wineries, mining and horse breeding) compete for land, water and space, presently condensed into and dubbed a 'foals vs. coals' scenario (Chapter 13). Yet in many other places local councils would welcome such intense competition and the employment that stems from it.

Australia has become more complex and more diverse – perhaps increasingly divided as a nation from north-west to south-east – symbolised by the unwillingness of unemployed workers to cross the Nullarbor for jobs. But, even more obviously, by a divide beyond which only sponge cities and mining centres survive, whereas in the penumbra of the metropolitan capitals small towns have survived, buoyed by counterurbanisation and tourism, but wholly changed from agricultural pasts. When Phil McManus and Bill Pritchard concluded *The Land of Discontent* (2000), at the start of the century, Cairns was singled out as a place of rapid transformation primarily through tourism and some degree of industrialisation. Since then growth has continued but centred on its port – for commodity exports – and airport, as Queensland's centre of FIFO. Typically growth has shifted inland, but the population has not. Indeed the image of mineworkers crowded into airports across the country, as they FIFO, is a potent symbol of the transience of success – how mineral booms can fade and mining towns fall off the map, as they have done for 200 years. The constants in rural and regional change are diversity, change and uncertainty.

References

Alston, M., 1995, *Women on the Land: The Hidden Heart of Rural Australia*. Sydney: UNSW Press.

Argent, N., 2011, 'What's New about Rural Governance? Australian Perspectives and Introduction to the Special Issue', *Australian Geographer*, 42(2): 95-103.

Argent, N. and Walmsley, J., 2008, 'Rural Youth Migration Trends in Australia: An Overview of Recent Trends and Two Inland Case Studies', *Geographical Research*, 46: 139-52.

Argent, N. and Rolley, F., 2000, 'Financial Exclusion in Rural and Remote New South Wales, Australia: A Geography of Bank Branch Rationalisation, 1981-98', *Australian Geographical Studies*, 38: 182-203.

Atherley, K., 2006, 'Sport, Localism and Social Capital in Rural Western Australia', *Geographical Research*, 44: 348-360.

Austin-Broos, D., 2009, *Arrernte Present, Arrernte Past: Invasion, Violence and Imagination in Indigenous Central Australia*, Chicago: University of Chicago Press.

Australian Government, 2012, *Feeding the Future*, Canberra: Department of Foreign Affairs and Trade.

Australian Labour Party, Windsor, A. and Oakeshott, R., 2010, 'Commitment to Regional Australia', in *Agreement for Government*, Canberra, Accessed 2 February 2013: http://resources.news.com.au/files/2010/09/07/1225915/542989-final-agreement-with-the-independents.pdf.

Beer, A., 2006, 'Regionalism and Economic Development: Achieving an Efficient framework', in Brown, A. and Bellamy, J. (eds), *Federalism and Regionalism in Australia: New Approaches, New Institutions?* Canberra: ANU E. Press, 119-34.

Beer, A., Maude, A. and Pritchard, B., 2003, *Developing Australia's Regions: Theory and Practice*, Sydney: UNSW Press.

Brett, J., 2007, 'The Country, the City and the State in the Australian Settlement', *Australian Journal of Political Science*, 42: 1-17.

Budge, T., Butt, A., Chesterfield, M., Kennedy, M., Buxton, M. and Tremain, D., 2012, *Does Australia need a National Policy to Preserve Agriculture Land?*, Sydney: Australian Farm Institute.

Burnley, I. and Murphy, P., 2004, *Sea Change. Movement from Metropolitan to Arcadian Australia*, Sydney: UNSW Press.

Carrington, K. and Pereira, M., 2011, 'Assessing the Social Impacts of the Resources Boom on Rural Communities', *Rural Society*, 21: 2-20.

Cheshire, L., 2006, *Governing Rural Development: Discourses and Practices of Self-Help in Australian Rural Policy*, Aldershot: Ashgate Publishing.

Cheshire, L., Higgins, V. and Lawrence, G. (eds), 2007, *Rural Governance: International Perspectives*, London: Routledge.

Cocklin, C. and Alston, M., 2003, *Community Sustainability in Rural Australia: A Question of Capital?* Wagga Wagga: Centre for Rural Social Research.

Connell, J., 2010, 'From Blackbirds to Guestworkers in the South Pacific: Plus ça change ...?', *Economic and Labour Relations Review*, 20: 111-122.

Connell, J. and McManus, P., 2011, *Rural Revival? Place Marketing, Tree Change and Regional Migration in Australia*, Farnham: Ashgate Publishing.

Danaher, M., 2008, 'Seeing the Change in a Sea Change Community: Issues for Environmental Managers', *Australasian Journal of Environmental Management*, 1: 51-60.

Davies, K., 2009, 'The Outback Gallery where Utes Rule', *The Sunday Telegraph*, 23 August, 6.

Dawson, A., 2012, 'Dark Side of the Boom', *About the House*, 45, August, 47-50.

Dufty, R. and Liu, E., 2011, '"Picking Blueberries and Indian women go Hand-in-hand": The Role of Gender and Ethnicity in the Division of Agricultural Labour in Woolgoolga, New South Wales, Australia', in Pini, B. and Leach, B. (eds), *Reshaping Gender and Class in Rural Spaces*, Farnham: Ashgate Publishing, 73-90.

Edwards, R., 2012, 'Gympie's Country Music Muster: Creating a Cultural Economy from a Local Tradition', *Journal of Rural Studies*, 28: 517-527.

Forth, G. and Howell, K., 2002, 'Don't Cry for Me Upper Wombat: The Realities of Regional/Small Town Decline in Non-Coastal Australia', *Sustaining Regions*, 2: 4-11.

Gibson, C. and Connell, J., 2012, *Music Festivals and Regional Development in Australia*, Farnham: Ashgate Publishing.

Gibson, C. and Davidson, D., 2004, 'Tamworth, Australia's "Country Music Capital": Place Marketing, Rurality, and Resident Reactions', *Journal of Rural Studies*, 20(4): 387-404.

Gorman-Murray, A., Waitt, G. and Gibson, C., 2012, 'Chilling Out in "Cosmopolitan Country": Urban/Rural Hybridity and the Construction of Daylesford as a "Lesbian and Gay Rural Idyll"', *Journal of Rural Studies*, 28: 69-79.

Gray, I., 1991, *Politics in Place: Social Power Relations in an Australian Country Town*, Cambridge: Cambridge University Press.

Gray, I. and Lawrence, G., 2001, *A Future for Regional Australia: Escaping Global Misfortune*, Cambridge, Cambridge University Press.

Green, D., Alexander, L., McInnes, K., Church, J., Nicholls, N. and White, N., 2010, 'An Assessment of Climate Change Impacts and Adaptation for the Torres Strait Islands, Australia', *Climatic Change*, 102: 405-433.

Guiney, R., 2012, 'Farming Suicides During the Victorian Drought: 2001-2007', *Australian Journal of Rural Health*, 20(1): 11-5.

Hart, C., Berry, H. and Tonna, A., 2011, 'Improving the Mental Health of Rural New South Wales Communities Facing Drought and Other Adversities', *Australian Journal of Rural Health*, 19: 231-238.

Herbert-Cheshire, L., 2000, 'Contemporary Strategies for Rural Community Development in Australia: A Governmentality Perpective', *Journal of Rural Studies*, 16: 203-215.

Herbert-Cheshire, L. and Higgins, V., 2004, 'From Risky to Responsible: Expert Knowledge and the Governing of Community-Led Rural Development', *Journal of Rural Studies*, 20: 289-302.

Herbert-Cheshire, L. and Lawrence, G., 2002, 'Political Economy and the Challenge of Governance', *Journal of Australian Political Economy*, 50: 139-145.

Higgins, V., 2001, 'Governing the Boundaries of Viability: Economic Expertise and the Production of the "Low-Income Farm Problem" in Australia', *Sociologia Ruralis*, 41(3): 358-375.

Holmes, J., 2002, 'Diversity and Change in Australia's Rangelands: A Post-Productivist Transition with a Difference?', *Transactions of the Institute of British Geographers*, 27: 362-384.

Holmes, J., 2006, 'Impulses Towards a Multifunctional Transition in Rural Australia: Gaps in the Research Agenda', *Journal of Rural Studies*, 22(2): 142-160.

Holmes, J., 2012, 'Cape York Peninsula, Australia: A Frontier Region Undergoing a Multifunctional Transition with Indigenous Engagement', *Journal of Rural Studies*, 28: 252-265.

Hugo, G., 2004, *A New Paradigm of International Migration: Implications for Migration Policy and Planning in Australia*, Canberra: Commonwealth of Australia.

Hugo, G. and Morén-Alegret, R., 2008, 'International Migration to Non-metropolitan Areas of High Income Countries: Editorial Introduction', *Population, Space and Place*, 14: 473-477.

Hugo, G. and Smailes, P., 1985, 'Urban-Rural Migration in Australia: A Process View of the Turnaround', *Journal of Rural Studies*, 1: 11-30.

Humphries, D., 2012, 'Good Times Reignite the Water Wars', *Sydney Morning Herald*, 2 June, 6-7.

Joint Select Committee on Migration, 2001, *New Faces, New Places: Review of State-specific Migration Mechanisms*, Canberra, Commonwealth of Australia.

Jordan, K., Krivokapic-Skoko, B. and Collins, J., 2009, 'The Ethnic Landscape of Rural Australia: Non-Anglo-Celtic Immigrant Communities and the Built Environment', *Journal of Rural Studies*, 25: 376-385.

Lawrence, G., 1987, *Capitalism and the Countryside: The Rural Crisis in Australia*, Sydney: Pluto Press.

Lawrence, G., Lyons, K. and Momtaz, S., 1996, *Social Change in Rural Australia*, Rockhampton: Central Queensland University Publishing.

Liepins, R., 2000, 'Making Men: The Construction and Representation of Agriculture-based Masculinities in Australia and New Zealand', *Rural Sociology*, 65(4): 605.

Liepins, R., 1998, 'The Gendering of Farming and Agricultural Politics: A Matter of Discourse and Power', *Australian Geographer*, 29(3): 371.

Lockie, S. and Bourke, L., 2001, *Rurality Bites: The Social and Environmental Transformation of Rural Australia*, Sydney: Pluto Press.

McManus, P., 2005, *Vortex Cities to Sustainable Cities: Australia's Urban Challenge*, Sydney: UNSW Press.

McManus, P., 2008, 'Their Grass is Greener but Ours is Sweeter – Thoroughbred Breeding and Water Management in the Upper Hunter Region of New South Wales, Australia', *Geoforum*, 39(3): 1296-1307.

McManus, P. and Pritchard, B., 2000, 'Concluding Thoughts', in P. McManus and B. Pritchard (eds), *Land of Discontent. The Dynamics of Change in Rural and Regional Australia*, Sydney: UNSW Press, 218-222.

McManus, P., Walmsley, J., Argent, N., Baum, S., Bourke, L., Martin, J., Pritchard, B. and Sorensen, A., 2012, 'Rural Community and Rural Resilience: What is Important to Farmers in Keeping their Country Towns Alive?', *Journal of Rural Studies*, 28: 20-29.

Martin, J., Budge, T. and Butt, A., 2011, 'Introduction', in J. Martin and T. Budge (eds), *The Sustainability of Australia's Country Towns: Renewal, Renaissance and Resilience*, Ballarat: Victorian Universities Regional Research Network Press, 1-10.

Martinez-Fernandez, C., Wu, C., Schatz, L., Taira, N. and Vargas-Hernandez, J., 2012, 'The Shrinking Mining City: Urban Dynamics and Contested Territory', *International Journal of Urban and Regional Research*, 36: 245-260.

Maude, A., 2003, 'Local and Regional Economic Development Organisations in Australia', in A. Beer, G. Haughton and A. Maude (eds), *Developing Locally: An International Comparison of Local and Regional Economic Development*, Bristol: Policy Press, 109-36.

Panelli, R., Hubbard, P., Coombes, B. and Suchet-Pearson, S., 2009, 'De-centring White Ruralities: Ethnic Diversity, Racialisation and Indigenous Countrysides', *Journal of Rural Studies*, 25: 355-364.

Pini, B., 1998, 'The Emerging Economic Rationalist Discourse on Women and Leadership in Australian Agriculture', *Rural Society*, 8(3): 223–233.

Plummer, P. and Tonts, M., 2013, 'Geographical Political Economy, Local Models, and the Evolution of a Resource Dependent Economy: The Pilbara Mining Region, 1980-2010', *Australian Geographer*, 44: in press.

Pritchard, B., 1999, 'Australia as the Supermarket to Asia? Governments, Territory and Political Economy in the Australian Agri-food System', *Rural Sociology*, 64(2): 284.

Pritchard, B., Argent, N., Baum, S., Bourke, L., Martin, J., McManus, P., Sorenson, A. and Walmsley, J., 2010, 'Local – If Possible: How the Spatial Networking of Economic Relations amongst Farm Enterprises Aids Small Town Survival in Rural Australia', *Regional Studies*, 46: 539-557.

Pritchard, B. and McManus, P. (eds), 2000, *Land of Discontent: The Dynamics of Change in Rural and Regional Australia*, Sydney: UNSW Press.

Sherval, M. and Greenwood, G., 2012, '"Drought-proofing" Regional Australia and the Rhetoric Surrounding Tillegra Dam, NSW', *Australian Geographer*, 43: 253-272.

Skeldon, R., 2006, 'Interlinkages between Internal and International Migration and Development in the Asian region', *Population, Space and Place*, 12: 15-30.

Spaaij, R., 2009, 'The Glue that Holds the Community Together? Sport and Sustainability in Rural Australia', *Sport in Society*, 12: 1132-1146.

Tonts, M. and Atherley, K., 2005, 'Rural Restructuring and the Changing Geography of Competitive Sport', *Australian Geographer*, 36: 125-144.

Tonts, M., Argent, N., Plummer, P., 2012, 'Evolutionary Perspectives on Rural Australia', *Geographical Research*, 50: 291-303.

Tonts, M., 2001, 'Rural Restructuring, Policy Change and Uneven Development in the Central Wheatbelt of Western Australia', *Social and Cultural Geography*, 2: 108.

Wheeler, S., Bjornlund, H., Zuo, A. and Edwards, J., 2012, 'Handing Down the Farm? The Increasing Uncertainty of Irrigated Farm Succession in Australia', *Journal of Rural Studies*, 28: 266-275.

Wulff, M., Carter, T., Vinebeg, R. and Ward, S., 2008, 'Attracting New Arrivals to Smaller Cities and Rural Communities – Findings from Australia, Canada and New Zealand', *International Journal of Migration and Integration*, 9: 119-124.

Chapter 2

The Role of Gender in the Migration Practices and Aspirations of Australian Rural Youth

Rae Dufty-Jones, Neil Argent, Fran Rolley and Jim Walmsley

Introduction

Youth out-migration, and youth net migration loss, have been prevailing trends in Australia's population geography since the early decades of the twentieth century (Hugo 2004). In the second decade of the twenty-first century this trend has become more entrenched, with it being almost *de rigueur* for many Australian youth to leave their rural towns and regions in search of new opportunities and experiences elsewhere (Argent and Walmsley 2008). In recent years there has been much public and academic debate around why rural youth, both in Australia and in other countries, have a higher propensity to migrate than their metropolitan counterparts and what this means for those communities they leave behind (Argent and Gibson 2008, Connell and McManus 2011, Stockdale 2006). Concern centres not just on the sheer loss of young people from demographically and economically struggling regions and communities but also on the extent to which they can be encouraged to return at some later stage of their life course and assume, *inter alia*, key local leadership and volunteer roles (Stockdale 2006). Gender has formed an important dimension in analyses of rural youth out-migration (Ni Laoire 1999, 2001, Alston 2004). However, the prevalence of return migration among young out-migrants, and the role of gender in influencing this process, has largely remained anecdotal. This chapter focuses on the contemporary rural population issue of youth out- and return- migration in Australia with a specific concern to understand how gender influences the 'cultures of migration' that inform decisions to leave rural locations and how gender might influence the return migration aspirations of rural youth out-migrants.

Rural Youth Migration Patterns in Twenty-first Century Australia

Australia has a high level of youth migration. As Argent and Walmsley (2008) noted, in the twenty years leading up to the 2001 Census, rates of migration for those aged between 15-24 years were among the highest of any age group in the nation. This trend appears to have continued into the twenty-first century with 49 percent of people aged 15-24 years identifying as having changed their place of residence in the five years leading up to 2011. This trend becomes even more pronounced (59 percent) when examining those aged between 20-24 years. Breaking this down into the crude regional categories of the major capital cities and the 'rest of' regions for each Australian State, those aged 15-24 years and living outside of the major metropolitan centres had a higher propensity to have changed their place of residence between 2006-2011 than their major capital city counterparts (52 percent compared to 47 percent). Furthermore, when we examine the migration patterns of 15-24 year olds in the 'rest of' Australian States by sex, there appear to be important gender differences, with 56 percent of young women aged between 15-24 years living outside of Sydney having moved at least once since 2006 compared to 49 percent of men from the same age cohort. While at a finer scale the patterns, processes, causes and impacts of rural youth migration are distributed in a far more spatially-uneven fashion (Argent and Walmsley 2008), the broad brush data presented from the 2011 Australian Census of Population and Housing suggests that rural youth migration, together with their gender aspects, continue to be significant factors influencing the social, economic and demographic future of rural Australia.

Rural Youth Out and Return Migration: Causes, Consequences and the Role of Gender

A number of factors have been identified for why rural youth are more likely to leave their home areas. The literature on rural youth migration identifies two broad sets of factors: structural and socio-cultural. Structural factors include the understanding that young people are more likely to move because of their 'life stage'. Billari (2001) refers to this period as the 'transition to adulthood', including milestones such as leaving school, commencing further education and full-time work, leaving the parental home, etc. However, this classic 'rite of passage' stage seems to occur earlier for youth in rural areas and often results in a move away from their home region because of the limited local availability of employment and education opportunities (Thissen *et al.* 2010). While education opportunities have always been comparatively limited in rural regions in Australia, the restructuring of Australian rural economies over the last forty years – which has resulted in a reduction in the number of employment opportunities in local labour markets, while at the same time increasing the need for higher educational qualifications in many sectors across the economy – has exacerbated this long-term trend (Alston 2004). Migration by rural youth is therefore an important mechanism for overcoming locational disadvantage, with migrants usually benefiting from their decisions to leave, despite initial challenges (Stockdale 2006, Argent and Walmsley 2008). The dual structural factors of economy and education point to an important class component in rural youth out-migration (Jones 1999, Jamieson 2000). The 'best and the brightest' use their human capital resources as a means of escaping the poor economic prospects that their home rural regions offer, leaving those less well-resourced with the challenge of remaining in the local employment market (Ni Laoire 2001, Gabriel 2002).

The rural youth migration literature has also been influenced by wider debates in the social sciences regarding the effect that an increasingly hyper-mobile, post-modern culture has on individual connections to community, family and place (Urry 2000, Bauman 2001). Accompanied by a shift towards a more 'biographical' approach to analysing migration, with migrants' identities being recognised as important factors shaping mobility (Boyle 2002), structural explanations of rural youth migration have been complemented by a focus on socio-cultural aspects. For example, some attribute rural youth out-migration to young people seeking to escape a kind of rural 'dull' (Rye 2006) or as a means of evading the surveillance of their activities by parents and communities (Ni Laoire 1999, Kraack and Kenway 2002, Alston, 2004). Easthope and Gabriel's (2008) concept of 'cultures of migration' further elaborates this idea by highlighting the role of peers, institutions and other important influences shaping young rural people's ideas and expectations concerning future careers and, accordingly, places to live. In relation to rural youth migration, the 'culture of migration' concept can be seen in the way in which the migration of young adults is normalised and expected. The normalisation process is often constructed through family and social networks. Parental and sibling migration biographies provide a migration template, and parents, teachers and friends establish expectations that many rural young people will one day leave. Such 'cultures of migration' only become more entrenched over time (Jones 1999, Stockdale 2002a, Bjarnason and Thorlindsson 2006, Connell 2008, Easthope and Gabriel 2008).

Does Rural Youth Out-Migration Matter?

A number of authors note the 'bleak' media representations of rural youth out-migration as one dimension of the manifold maladies affecting rural and regional Australia (Gabriel 2002, Argent and Walmsley 2008, Easthope and Gabriel 2008). Indeed, the consequences are severe for those communities that experience prolonged rural youth out-migration. First is the erosion of critical social capital, including important social and economic members of rural communities, be they contributing to the community in a leadership and/or business sense or as members of local clubs (Tonts and Atherley 2005, Bjarnason and Thorlindsson 2006). Furthermore, the loss of a significant proportion of young people from rural areas drives the structural ageing of rural communities. As a consequence, a diminishing youth population can prompt a vicious cycle of decline for rural communities, with lower population numbers overall affecting the sustainability of community services and businesses.

Despite these consequences, policy responses to rural youth out-migration that simply seek to retain young people in rural communities have begun to be questioned (Eversole 2001, Stockdale 2004, 2006, Argent and Walmsley 2008). Stockdale (2006: 354) argues that rural youth out-migration is an important part of rural economic regeneration as it is only 'by leaving rural areas [that] young adults can acquire the necessary skills to participate in endogenous development'. Instead, the problem of rural youth out-migration needs to be reframed so that the focus is on the level of return migration and/ or in-migration of younger residents to the same communities. In terms of return migration of former youth residents a number of structural and socio-cultural barriers to returning have been identified, including the availability of suitable employment, a return being equated with 'failure', concerns over the loss of social and other lifestyle opportunities, and of personal freedom, and the general 'culture shock' many experience when they do return (Stockdale 2006, Ni Laoire 2007, Davies 2008). These barriers often outweigh the key 'pull' factors that inform a desire to return.

Where Does Gender Fit in Rural Youth Migration?

Since the early 1990s, migration research has become more concerned with how gender influences migration decisions and the impacts of this (Bonney and Love 1991, Boyle and Halfacree 1995, Halfacree and Boyle 1999). Silvey (2004: 491) identifies a 'gender division of mobility' that influences both internal and international migration flows. She argues that this research not only explores the ways that gender shapes migration but also the way that migration shapes gendered 'social orders, geographies of inequality, spatialised subjectivities and the meanings of difference' (Silvey 2004: 491). Indeed, the gender dimension of rural youth out-migration, in terms of its propensity, causes and consequences, has been recognised by a number of scholars (Ravenstein 1885, Ni Laoire 1999, Jamieson 2000, Ni Laoire 2001, Alston 2004). Bjarnason and Thorlindsson (2006: 298) argue that the pattern of rural women migrating to metropolitan centres is 'one of the most consistent findings of migration research'. Research has found that young rural women are more likely to anticipate the need to leave (Bjarnason and Thorlindsson 2006) and are more likely to leave (Hugo 2004, Argent and Walmsley 2008).

Explanations for why gender plays such an important role in rural youth out-migration also follow the structural and socio-cultural themes that inform analyses of rural youth migration generally. Structural factors that influence female youth out-migration from rural areas include the more limited and/ or low-paid employment prospects that many young rural women face after leaving school (Bjarnason and Thorlindsson 2006, Corbett 2007). In many rural areas labour markets remain strongly sex-segregated, exacerbating female out-migration as many young women pursue higher education at city universities (Ni Laoire, 1999). The link between local employment opportunities and the out-migration of young Australian rural women in pursuit of educational opportunities has been explored by James (in Hare 2011: 8):

> Since the 1970s, two-thirds of all jobs for girls have disappeared [from rural regions]... Only half the jobs for boys have gone. The full-time job market for young people has been held together by apprenticeships, to which girls have very little access... They [girls] have no choice... It doesn't mean their experience of school is any better than boys, that they are happier or they are making the right choices. What it means is that the labour market won't accept them and is forcing them to aim for higher levels of the labour market which are credentials-based.

Intertwined with the structural push factors of limited local employment prospects and education opportunities is the role of socio-cultural factors. For instance, younger rural women may be put off remaining in their home town by the informal social expectations that place rural women in the role of performing substantial unpaid family and community work (Bjarnason and Thorlindsson 2006). Or they may view out-migration as a means of escaping a 'macho' culture, where, in many rural communities, public spaces (e.g. pubs, sports venues) are appropriated by young males for masculine performances (Ni Laoire 1999, Kraack and Kenway 2002, Alston 2004). Similarly, the strength and durability of the rural idyll is argued to reinforce gendered divisions of labour and domesticity for women (Hunter and Riney-Kehrberg 2002: 135). For example, labour force divisions that construct apprenticeships and careers in the primary or secondary rural industries as predominantly for men can be reinforced through parental expectations, with sons preferred in terms of who takes over the family farm (Ni Laoire 1999). The difference in rural parental expectations of sons and daughters is evidenced in a recent report by Baxter *et al.* (2011: 5) which found that in outer regional areas 59 percent of parents expected their daughter to obtain a university-level qualification compared to 40 percent having the same expectation for sons. Such gendered education and career expectations directly feed into cultures of rural youth migration. However, as Ni Laoire (1999, 2001) points out, the enablement of young men to remain behind in rural communities does not necessarily mean that they are in a privileged position. Those who remain behind are exposed to the stresses and strains of rural restructuring, the lack of qualifications and traditional forms of masculine identity that contribute to feelings of entrapment and hopelessness. Out-migration also reduces the social support networks available to men who remain behind, resulting in higher levels of stress and rising suicide rates.

While much headway has been made both in accounting for gender generally in migration research and specifically in rural youth migration, little is known about how gender influences the cultures of migration that inform rural youth out-migration decisions and how gender frames the return migration aspirations of those who leave.

Rural Youth Out- and Return Migration, Armidale, NSW

Current research on rural youth migration is constrained by two factors: first, census data provides only a snapshot on the processes and patterns of rural youth migration and gives no insight into the motivations behind migration decisions (Gibson and Argent 2008); second, where research is more qualitative, there are a number of difficulties in tracing out-migrants once they have left their home community (Stockdale 2002b). In response to these challenges, this research adopted a biographical approach to the collection and analysis of the data (Halfacree and Boyle 1993). As Halfacree and Boyle (1999: 8) explain, a biographical approach recognises that processes of migration are a product of both 'responsible action and ... occur within the messy "hurly-burly" of everyday life'. Employing this framework in migration research involves taking a more longitudinal perspective, assuming that actors are not always 'rational' and that migration decisions are entangled in the multiple concerns of both individuals and households at any one time. A biographical approach therefore is a useful framework for shedding light on the complexities of rural youth migration generally and the gender dimensions of this process specifically.

The research employed a mixed-method approach of an online questionnaire (distributed primarily through the online school and year related social networks on Facebook) and follow-up interviews with those who identified as being available for an interview in the online questionnaire. The research focused on two regions in Northern New South Wales (NSW), Australia: the Northern Tablelands and the Mid-North Coast and targeted two cohorts of rural youth, those aged 15-16 years (therefore most likely to be in Year 10) in the Census years of 2001 and 2006. At the time of interviewing (2011-12) these young people were aged 25-26 years and 20-21 years respectively. These former students were invited to participate in the research if they had attended high schools within case study region towns: Armidale, Uralla and Barraba (Northern Tablelands), and Coffs Harbour, Bellingen and Nambucca Heads (Mid-North Coast).

Befitting this generation's apparent preference for online media as the dominant method of making and maintaining social relationships, the research was conducted primarily via online means. Facebook was used to trace, contact and recruit former students and proved particularly useful, but also supplemented with the more traditional method of the 'phone book'.

The questionnaire sought details of individuals' migration histories, reasons for moving, satisfaction with home and new community life, and migration intentions. A total of 992 former students across both case study areas were identified and invited to participate in the research, with 135 completed surveys returned (a 15 percent response rate). Although not a high response rate it should be borne in mind that this was effectively a survey of a *population* (e.g. all former students in 2001 and 2006). Of those who responded, 76 percent were 'movers' (i.e. living in a different place of permanent residence than

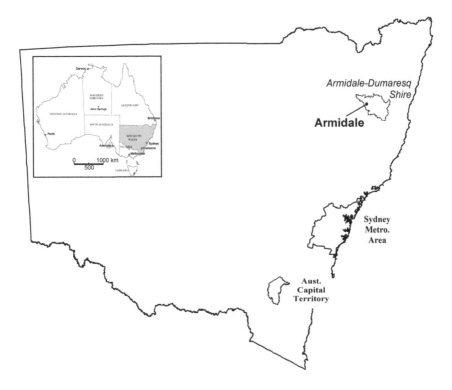

Figure 2.1   Location of Armidale and Sydney, NSW, Australia

where they studied), 11 percent were 'stayers' (i.e. living in the same town where they studied) and 13 percent were 'returnees' (i.e. now living in the same town as they did when studying after having lived somewhere else).

The following analysis and discussion focuses on respondents from Armidale (see Figure 2.1). Twenty-seven former Armidale students who completed the questionnaire volunteered to be interviewed. In-depth interviews provided deeper insights into the complexity of the (im)/mobility choices made by respondents (Stockdale 2002b). Gender emerged from the interviews as an important factor informing the mobility choices of out-migrants both in terms of their decisions to move as well as their aspirations to return. The remainder of this chapter draws primarily on a series of in-depth interviews with former students once resident in the regional city of Armidale but now living in Sydney (n=10; 6 female and 4 male). Gender was seen to be influential in these young people's migration decision-making in two key aspects: first, in terms of who and what influenced the decision to leave; and, second, in terms of future migration aspirations, especially the possibility of return migration.

The Role of Gender in Making the Decision to Leave

As previously noted, the role of families and social networks in influencing young rural people to make the decision to leave is a long-standing feature of the rural youth migration literature (Jones 1999, Jamieson 2000, Stockdale 2002a, Bjarnason and Thorlindsson 2006, Drozdzewski 2008, Easthope and Gabriel 2008). This also proved to be true for the former Armidale students living in Sydney. However, unlike previous studies, the current research found that there were distinct gender dimensions to these influences, especially for women, who identified strong female role models who played both positive and negative roles in influencing their decision to leave Armidale.

Consistent with the findings of Jones (1999: 18) who argued that 'family history of migration and rootedness may variously affect migration', this research found that female family members, in particular mothers, performed a positive role in terms of why young Armidale women eventually made the decision to leave. In some instances a mother's own migration biography formed an important template that informed the migration decisions and experiences of some of the young women interviewed. As one young female out-migrant from Armidale explained:

> Mum left home when she was 18, went from Grafton to Armidale, which back then was a big deal. Her parents didn't have a phone. So it was just like if mum could do it [then I could too] ... (female, 2001, Armidale/Marrickville)

Another respondent provided a similar explanation for the role that her mother played in influencing her decision to migrate: '... mum had done that [moved away from her home rural town] to study. I think, knowing that, I was "oh yeah, well mum did it"' (female, 2001, Armidale/Newtown). For this respondent her mother had not only provided a 'migration template' to build a certain culture of migration from, but had also actively encouraged her daughter to consider moving away to pursue a career in design:

> It's funny because mum always used to say – I'd always been drawing and quite design minded – she was always saying, you'll go to Sydney one day. You'll be off to Sydney. I think that it [the idea of moving away from Armidale] probably just clicked. She knew it before I did that that's what I'd have to do. She was completely right... (female, 2001, Armidale/Newtown).

Mothers therefore performed a particularly important role in establishing and maintaining a culture of migration for some of the interviewees. Mothers often played a dual role in providing a type of 'migration biography template' from which interviewees could draw on as part of envisaging and then justifying their own decision to move, particularly when times got tough; and in directly planting the seed of the idea of migration in respondents' minds from an early age. Interestingly, not one of the male interviewees referred to this as an important reason for leaving.

Other female family members were also influential in the decision to migrate away from Armidale. Siblings, particularly sisters, were important not only in providing a 'migration template' to model from, but also contributing important resources to facilitate the process of out-migration.

> It kind of happened by accident. My sister is six years older than me. She moved here [Sydney] after she finished uni. So I came for a visit, to help one of her friends do a play that they were putting on. The director of the play had a job going at a jewellery place that my sister worked at. So I kind of said 'oh yes, I'll do that until the play ends', and then I'll go back to Armidale. Then it was just like well, I might just live here actually because I don't want to [go back to Armidale] ... (female, 2001, Armidale/Chippendale).

In this instance, the interviewee's sister provided important economic resources in terms of housing and job networks that significantly lessened the costs of migration and assisted the interviewee to eventually make the decision to permanently live in Sydney.

The gendered nature of who influenced individuals to leave Armidale was also evident in the way in which a number of young female out-migrants used what they saw as the negative examples of those young women who remained behind in Armidale as a key reason for why they decided to leave. This research found class distinctions between movers and stayers to be woven into wider justifications for why some out-migrants had made the decision to leave. However, these class distinctions were further nuanced for females in terms of a recurring account of what happened to those young rural women who remained behind in Armidale. For some of the 'movers' interviewed, young women who remained behind were seen as limiting their own career prospects, due to 'early' marriage and child-bearing.

> 'That's what happens when you stay in Armidale'. That's a big catch phrase: 'that's what happens when you stay in Armidale'; you have babies and you get married [laughs]; as if it's the worst thing in the world (female, 2001, Armidale/ Chippendale).

Another respondent, who reflected on the experiences of her female contemporaries remaining in Armidale, provided a more personal account of the same consequences:

> All these girls that I went to school with, a lot of them are in Armidale. I think probably nine out of 10 of them have kids now. ... I find that amazing. I'm all for kids. It's just ... I just can't understand how – I just don't know how I would be okay with that, in what situation? I mean, maybe, in a small way, you envy that. They're totally content and they're happy (female, 2001, Armidale/ Newtown).

Both respondents temper their judgements of 'she who remains behind', acknowledging that 'getting married and having babies' is not 'the worst thing in the world' and that those who chose this path were most probably 'totally content and ... happy'. Other interviewees were more condemnatory in their assessment of Armidale 'stayers' and the life course that they apparently naturally follow:

> I didn't want to go there [University of New England], one, because – just seeing people who stay in Armidale, they're generally a certain type of person and I didn't really want to – not people who go to UNE from other places, but people who actually stay in Armidale. You get in on a 50 [University Admission Index] and stay friends with the same people that you were friends with in high school and just then kind of stagnate until you get married and have kids (female, 2001, Armidale/Marrickville).

Other authors have also noted the way in which out-migrants often defined themselves in opposition to those who remained behind (Jones 1999, Jamieson 2000, Easthope and Gabriel 2008, Thissen *et al.* 2010). Such dichotomous constructions are argued to be based around class differences where those 'with finer qualities normally get out leaving behind those stuck in their narrow mind set' (Jamieson 2000: 217-8). While Cuervo and Wyn (2012) caution that there may be little reality to the dichotomous construction of urban-positive/rural-negative and that the experiences of urban and rural young people are more similar than these accounts present, the above analysis does demonstrate how powerful and pervasive both class and gendered judgements of young women 'stayers' are for young female out-migrants. Reflecting on the consequences for 'she who remains behind' was important in both informing initial decisions to leave and as a means of justifying out-migration. Importantly, however, such considerations and influences on the cultures of migration had no relevance for the young rural men interviewed.

Gender and Return Migration Aspirations

As outlined earlier, the social construction of rural youth out-migration as a 'problem' overlooks the important human capital resources that young rural people develop through education and work experiences. Rather, the real issue for rural communities is that those who leave to develop improved human capital resources often fail to return to where they could play an integral and informed role in regional development (Stockdale 2006). With few exceptions return migration has been an under-researched area in rural population studies with no studies extensively examining the gender dimensions of return rural migration. Yet, as this research shows, gender is an important factor when examining whether and how young rural out-migrants envisage a possible return.

Interviews with former Armidale school students currently living in Sydney showed that not only was there a culture of *out*-migration but that there was also a culture of *return* migration. As one respondent explained:

> ... when we first moved to Armidale, a lot of people said, 'oh you know, people always come back to Armidale'. There were a lot of people, a lot of parents of my friends who grew up in Armidale when we were there. So they had gone to Sydney or gone to Melbourne or whatever and done their thing and then come back to Armidale to settle (female, 2001, Armidale/Annandale).

Indeed, it is important to note that all interviewees expressed an aspiration to return to a rural location – if not Armidale specifically – at some point in the future. Some were surprised by their desire to leave Sydney: 'I'm sitting in Sydney to find the transition between uni and a career. Once I do that ... I'll probably try and go up north, because I'm not really a city person. That shocked me' (female, 2001, Armidale/Chippendale).

Gender was an important factor in these aspirations. For the female respondents, particularly those in committed relationships, the intention to return and the actual possibility of returning were two quite different things. Their desire to return to a rural location was countered by the equally strong need to maintain the careers they had established in Sydney:

> ... I am struggling with that [the idea of returning to a rural location]. I've thought about that because I really feel in a year's time, I could try and say, 'yeah, I can move back to the country'. ... but my job's here ... and I love my job and I don't want to leave it. What do I do then? Yeah, I don't know how it's going to work, particularly in my profession [interior design]. I think there are a lot of professions where you have to build a reputation in order for people to come to you. Then you might be in a position to say okay, well, I'm going to live here. I've thought about how I might tackle that one day. You live in the city but you have a place in the country that you go to and you have that dual lifestyle. It's a very hard one. I don't really know how it's going to happen but I do know that I definitely don't want to live in the city forever (female, 2001, Armidale/Newtown).

The same interviewee speculated that if she had trained for a more traditionally 'female' career she would have been better able to have a career, relationship and the choice of returning to the Northern Tablelands region:

> Yeah. I wonder if I did nursing or something, like a couple of friends of mine did, whether I'd be in Armidale to do that? I've got a really good friend who did nursing in Armidale and she's now in Tamworth. She's been working there for about a year, I think. She lives in Tamworth with her boyfriend (female, 2001, Armidale/Newtown).

Another female interviewee expressed a similar struggle. This struggle, however, was strongly contrasted with how her partner (also originally from Armidale) envisaged his own return migration:

> (male): But long-term, definitely move back to, probably, Armidale.

> (female): But see, long-term for me is not necessarily moving back to Armidale. It's just this utopian idea of living somewhere with a backyard so that I can have a dog ... I just feel – when I was younger, because we grew up in the country, so we had a dog. I want to have that life again where you have a backyard, you can finish work at 5 o'clock and be home by 5:30 so you can take your dog for a walk when it's still light... You know? Have maybe a backyard so that I can grow my own organic vegetables and stuff. It's a utopian [dream] – yeah.

> (male): I want a shed. For power tools.

> (female): Yeah, you want a man shed. ... [But it is hard to try and leave Sydney] Especially when you do have a career, which both of us do. So if I move back to a country town, even if I'm still working in public health, ... a change [a move back to Armidale] would mean changing my career path (female and male, 2001, Armidale/Marrickville).

In this interview the male interviewee was far more certain about his prospects of maintaining his career in aviation while also being able to return to live in Armidale. His partner, however, viewed her aspiration of returning to Armidale (with all its perks of affordable housing and good commuting times) and maintaining her career in medical research as 'utopian'. Her childhood experiences of growing up in a rural location were clearly positive and extremely influential in her desire to return to the country. However, the conflict between career and a rural lifestyle meant that these return migration aspirations were pared back to being able to afford a place with a backyard.

Consistent with Ni Laoire's (2007) findings, the desire to return to a rural location was also strongly influenced by a belief that it was a better location than a city to raise children. Again, both men and women identified this as an important aspect of why they would seek to return to Armidale specifically, or a rural location more generally:

> (male) I'd like to move back to the county purely as a sentimental thing. It's where you'd want your kids to grow up; it's where you want them to go to school – that sort of thing. ... Yeah, I was just talking last night – that was a catch up with the guys from TAS [The Armidale School – an elite private school located in Armidale] and we all share the same thing. We don't really want to live in the city. Maybe on the coast or back home.

Facilitator: What are the reasons for that? Why wouldn't you want to live in the city as opposed to – and it would be back home, it wouldn't be anywhere else?

(male): For me it might be back home. … Mostly it's just a sentimental thing; it's where I grew up… Just more probably like I own that, I feel responsible and I want to make [Armidale] a better place. This, it's just – … I don't own anything here [Sydney]. I'm not responsible for it (male, 2006, Armidale/Camperdown).

This interview with a male respondent indicates that, along with family, strong ties to community were also important factors in influencing a desire to make a future return to Armidale. The interviewee's sense of 'ownership' and 'responsibility' to make Armidale a 'better place' echoes other rural research (Shortall 1999) that highlights the strongly gendered nature of concepts of rural property and ownership. This perspective sits in contrast to the way in which family and the desire to return to live in a rural location were expressed by a female respondent:

(female): I suppose you get to a point where you think you can't do it [live in the city] forever. … You also … hit that age where you – and I'm not a traditionalist in any sense, but you do get to a point where you …

Facilitator: There's a biological reality that you have to deal with?

(female): Yeah, exactly. My mum's a midwife and she reminds me of that quite frequently. Just thinking about that and, yeah, family's the big thing, having kids, but also what your partner's doing. I don't know. It's hard that one. … I just don't know. I'm quite happy to not know where I'll be, but I feel like it always has to be a bit of a compromise which is mine. Yeah, I think I'd really struggle. I'd hate to have kids and not have the family around. In saying that, I think well, I've lived in Sydney for five years and I might go home three times a year – seeing my family four or five times a year. We talk every week, and I've become used to that. I think if I were to have kids in the city, you could become used to that but I don't want to (female, 2001, Armidale/Newtown).

Like the male interviewee earlier, this female respondent also identified the importance of family and living in a rural location. However, for this respondent, both because she is older than the male interviewee and because she is female, the biological pressure to make a decision about 'where' to have her family was felt to be mounting and was a far more immediate concern. Her desire to return to Armidale, or, at the very least, the country, was considerably more circumspect and counterbalanced by statements such as 'well, I've lived in Sydney for five years' and 'if I were to have kids in the city, you could become used to that'. That is, there is a *time* in life when return migration is viewed as more likely and possible. However, also significant in this excerpt are the unequal gender relations in making the decision of where a household should live, with the interviewee noting that she feels like 'it [the decision of where to live] always has to be a bit of a compromise which is mine'.

Conclusions: Does Gender Matter when it Comes to Rural Youth Out- and Return Migration in Australia?

This chapter has focused on rural youth out- and return migration in Australia and specifically how gender influences 'cultures of migration' that inform decisions to leave or return to rural locations. In terms of the cultures of migration that female interviewees identified, the influence of both positive and negative female role models was a particularly strong factor in informing the cultures of migration for these interviewees. Whether it was the positive female roles of mothers and sisters establishing cultures of migration or providing important economic and psychic resources in the actual process of migration or the negative narratives around 'she who remained behind' in Armidale, gender was a deeply influential dimension when understanding the cultures of migration that inform and are then used to justify why young rural women choose to leave their home towns.

Gender also informed the return migration aspirations of former rural out-migrants. While some interviewees identified a culture of return migration and all of them thus indicated a desire to eventually live in a rural location, if not Armidale itself, such aspirations were tempered by gender. For many female respondents, the potential changes to their careers that a return to Armidale or another rural location would entail were identified as a significant barrier to their general desire to return. For the young men interviewed, the potential career challenges that return migration might involve did not seem as insurmountable. This difference may reflect gendered cultures of migration that, for instance, remind young rural women that career prospects in rural locations are limited and that out-migration for education and employment is the surest way to ensure a meaningful career. Gender also tempered the way in which returning for 'family' reasons was presented by male and female respondents. Male respondents were more unequivocal about the idea of return migration for family; these discourses also had the added dimension of young men seeking to return to Armidale because they felt a sense of 'ownership' and 'responsibility' for the town and its community. This was juxtaposed by female accounts of their return migration aspirations as being 'utopian', a 'struggle', and a 'compromise'.

Ultimately it needs to be asked whether the gender aspects of cultures of migration and return migration aspirations matter. This study suggests that it does. The negative narrative of what happens to young rural women who remain behind is problematic because it reinforces a class-based and harmful stereotype of Australian rural locations as 'backwaters'. Furthermore, these narratives reinforce fears that can potentially prevent young rural women viewing return migration as a positive move for both career and family. The gender dimensions influencing return migration aspirations are also problematic. Australian rural communities have a lot to offer young returnees, including affordable housing, good commuting times and, therefore, work-life balance. Both men and women were aware of these benefits, however the young women interviewed were deeply sceptical of their ability to maintain a fulfilling career, if they make the decision to return, unlike their male counterparts. For many women, out-migration represents moving away from traditional gender-specific roles such as having children. Return migration may well represent a particularly gendered form of 'failure' for these rural women. Rural development strategies, particularly those focused around endogenous strategies, therefore face a considerable barrier convincing young female out-migrants that returning will be beneficial both in terms of lifestyle and career opportunities. Australian rural communities ignore this reality at their peril. By not trying to engage with and correct the fears that potential female returnees have about their employment prospects, rural communities are potentially missing out on a significant proportion of the human capital that is generated with out-migration.

References

Alston, M., 2004, '"You Don't Want to be a Check-out Chick all Your Life": The Out-migration of Young People from Australia's Small Rural Towns', *The Australian Journal of Social Issues*, 39: 299-313.

Argent, N. and Walmsley, J., 2008, 'Rural Youth Migration Trends in Australia: An Overview of Recent Trends and Two Inland Case Studies', *Geographical Research*, 46: 139-152.

Bauman, Z., 2001, *The Individualized Society*, Malden, MA: Polity Press.

Baxter, J., Gray, M. and Hayes, A., 2011, *Families in Regional, Rural and Remote Australia*, Melbourne: Australian Institute of Family Studies.

Billari, F., 2001, 'The Analysis of Early Life Courses: Complex Descriptions of the Transition to Adulthood', *Journal of Population Research*, 18: 119-142.

Bjarnason, T. and Thorlindsson, T., 2006, 'Should I Stay or Should I Go? Migration Expectations among Youth in Icelandic Fishing and Farming Communities', *Journal of Rural Studies*, 22: 290-300.

Bonney, N. and Love, J., 1991, 'Gender and Migrations: Geographical Mobility and the Wife's Sacrifice', *Sociological Review*, 39: 335-348.

Boyle, P., 2002, 'Population Geography: Transnational Women on the Move', *Progress in Human Geography*, 26: 531-543.

Boyle, P. and Halfacree, K., 1995, 'Service Class Migration in England and Wales, 1980-1981: Identifying Gender-Specific Mobility Patterns', *Regional Studies*, 29: 43-57.

Connell, J. and McManus, P., 2011, *Rural Revival? Place Marketing, Tree Change and Regional Migration in Australia*, Farnham: Ashgate Publishing.

Connell, J. 2008, 'Niue: Embracing a Culture of Migration', *Journal of Ethnic and Migration Studies*, 34: 1021-1040.

Corbett, M., 2007, 'All Kinds of Potential: Women and Out-migration in an Atlantic Canadian Coastal Community', *Journal of Rural Studies*, 23: 430-442.

Cuervo, H. and Wyn, J., 2012, *Young People Making it Work: Continuity and Change in Rural Places*, Carlton: Melbourne University Press.

Davies, A., 2008, 'Declining Youth In-migration in Rural Western Australia: The Role of Perceptions of Rural Employment and Lifestyle Opportunities', *Geographical Research*, 46: 162-171.

Drozdzewski, D., 2008, '"We're moving out": Youth Out-Migration Intentions in Coastal Non-Metropolitan New South Wales', *Geographical Research*, 46: 153-161.

Easthope, H. and Gabriel, M., 2008, 'Turbulent Lives: Exploring the Cultural Meaning of Regional Youth Migration', *Geographical Research*, 46: 172-182.

Eversole, R., 2001, 'Keeping Youth in Communities: Education and Out-migration in the South West', *Rural Society*, 11: 85-98.

Gabriel, M., 2002, 'Australia's Regional Youth Exodus', *Journal of Rural Studies*, 18: 209-212.

Gibson, C. and Argent, N., 2008, 'Getting On, Getting Up and Getting Out? Broadening Perspectives on Rural Youth Migration', *Geographical Research*, 46: 135-138.

Halfacree, K. and Boyle, P., 1993, 'The Challenge Facing Migration Research: The Case for a Biographical Approach', *Progress in Human Geography*, 17: 333-348.

Halfacree, K. and Boyle, P., 1999, 'Introduction: Gender and Migration in Developed Countries', in P. Boyle and K. Halfacree (eds), *Migration and Gender in the Developed World*, London: Routledge, 1-29.

Hare, J., 2011, 'School Ties Traded for King Gees', *The Weekend Australian*, 30 August, 14.

Hugo, G., 2004, 'The State of Rural Populations', in C. Cocklin and J. Dibden (eds), *Sustainability and Change in Rural Australia*, Sydney: UNSW Press, 56-79.

Hunter, K. and Riney-Kehrberg, P., 2002, 'Rural Daughters in Australia, New Zealand and the United States: An Historical Perspective', *Journal of Rural Studies*, 18: 135-143.

Jamieson, L., 2000, 'Migration, Place and Class: Youth in a Rural Area', *Sociological Review*, 48: 203-224.

Jones, G., 1999, '"The Same People in the Same Places"? Socio-Spatial Identities and Migration in Youth', *Sociology*, 33: 1-22.

Kraack, A. and Kenway, J., 2002, 'Place, Time and Stigmatised Youthful Identities: Bad Boys in Paradise', *Journal of Rural Studies*, 18: 145-155.

Ni Laoire, C., 1999, 'Gender Issues in Irish Rural Out-migration', in P. Boyle and K. Halfacree (eds), *Migration and Gender in the Developed World*, London: Routledge, 223-237.

Ni Laoire, C., 2001, 'A Matter of Life and Death? Men, Masculinities and Staying "Behind" in Rural Ireland', *Sociologia Ruralis*, 41: 220-236.

Ni Laoire, C., 2007, 'The "Green Green Grass of Home"? Return Migration to Rural Ireland', *Journal of Rural Studies*, 23: 332-344.

Ravenstein, E., 1885, 'The Laws of Migration', *Journal of the Statistical Society*, 48: 167-227.

Rye, J.F., 2006, 'Rural Youths' Images of the Rural', *Journal of Rural Studies*, 22: 409-421.

Shortall, S., 1999, *Women and Farming: Property and Power*, Basingstoke: Macmillan Press.

Silvey, R., 2004, 'Power, Difference and Mobility: Feminist Advances in Migration Studies', *Progress in Human Geography*, 28; 490-506.

Stockdale, A., 2002a, 'Out-migration from Rural Scotland: The Importance of Family and Social Networks', *Sociologia Ruralis*, 42: 41-64.

Stockdale, A., 2002b, 'Towards a Typology of Out-migration from Peripheral Areas: A Scottish Case Study', *International Journal of Population Geography*, 8: 345-364.

Stockdale, A., 2004, 'Rural Out-Migration: Community Consequences and Individual Migrant Experiences', *Sociologia Ruralis*, 44: 167-194.

Stockdale, A., 2006, 'Migration: Pre-requisite for Rural Economic Regeneration?', *Journal of Rural Studies*, 22: 354-366.

Thissen, F., Fortuijn, J.D., Strijker, D. and Haartsen, T., 2010, 'Migration Intentions of Rural Youth in the Westhoek, Flanders, Belgium and the Veenkolonien, The Netherlands', *Journal of Rural Studies*, 26: 428-436.

Tonts, M. and Atherley, K., 2005, 'Rural Restructuring and the Changing Geography of Competitive Sport', *Australian Geographer*, 36: 125-144.

Urry, J., 2000, *Sociology Beyond Societies: Mobilities for the Twenty-first Century*, London: Routledge.

Chapter 3

Urban to Rural Elderly Migration: Renewing and Reinventing Australia's Small Rural Towns

Amanda Davies

Introduction

As Australia's baby boomer population enters 'seniordom' they are increasingly consuming and producing new rural spaces. Australia has a particularly mobile older population, with many people choosing to move at, or shortly after, retirement (Davies and James 2011). Rural destinations, and in particular small towns, are high on the list of sought after retirement localities. This movement of older people into 'desired' rural localities is reshaping the social, economic and physical character of rural towns (Connell and McManus 2011). For some towns, in-migration of older people has underpinned a new phase of growth and socio-economic transformation, with vibrant new economic activities and social cultures (Davies 2009). Others, however, view this population growth in somewhat more negative terms, citing the upward pressure on housing prices, congestion, increased burden on local health care resources and a loss of the traditional character of places (Costello 2007, 2009, Gurran 2008, Ragusa 2010). This chapter looks at the reasons why older people are moving to rural Australia, what areas they are moving to and, how this movement is underpinning changes in the character, function and form of some of Australia's small rural towns.

Understanding Elderly Migration to Rural Areas

Australia's population is ageing in a manner comparable to that of the United Kingdom and the advanced industrial countries of Europe and North America (Kinsella and He 2009). In Australia, this demographic trend is highly spatially variable, with rural areas, on average, having a larger share of older people and experiencing population ageing more rapidly than metropolitan areas (Davies and James 2011). Moreover, rural areas fringing metropolitan regions tend to have older populations than more remote regions.

While there are some exceptions, most Australian rural towns have larger shares of people aged 65+ and smaller shares of people aged 15-35 when compared to metropolitan areas (Davies and James 2011). Over the last two decades, this share of older people to younger people in Australia's population has grown and again this trend has been more pronounced in rural towns (Davies and James 2011). Given that rural Australia has fewer healthcare and social support resources per capita, the above linked trends have given rise to claims that rural Australia is facing a 'population ageing crisis' (National Rural Health Alliance 2009, National Seniors Australia Productive Ageing Centre 2010, Spies-Butcher 2011). Thus the Productivity Commission (2005: xiii) commented,

> Population ageing has been called the quiet transformation, because it is gradual, but also unremitting and ultimately pervasive. Population ageing will accelerate over the next few decades in Australia, with far-reaching economic implications. It will slow Australia's workforce and economic growth, at the very time that burgeoning demands are placed on Australia's health and aged care systems.

On ageing in rural Australia, a study by KPMG (2012: 1) revealed,

> ...the family farm could be disappearing of its own accord. By some estimates, up to half of Australia's current population of farmers could retire during the coming decade. The average age of Australian farmers in 2011 was 56, so for many retirement will be a necessity, not an option. It is unclear where the replacements will come from.

While broadly speaking rural Australia's population is ageing, there is great variability in population ageing across rural Australia (Davies and James 2011). This spatial variability has, to a large extent, been caused by the migration of people between urban and rural spaces and also within rural spaces (Davies and James 2011). The most notable migration flows influencing the ageing of Australia's rural regions are the movement of young people from rural areas to large cities, primarily to access education, employment and diverse lifestyle opportunities; the movement of older people, at or near retirement, from metropolitan regions to rural localities of high natural amenity and in particular coastal areas within 300 kilometers of major cities; and the movement of elderly people away from very small rural settlements (less than 1000 people) to medium sized rural towns and regional cities.

Since the mid 1990s, there has been a marked increase in the number of older people moving to rural areas for their retirement (Davies and James 2011). Sander (2010) found that between 2001 and 2006 approximately 650,000 Australians aged 55-69 moved home. This represented 29 percent of all people aged 55-69. Of those who moved, 56 percent remained within the same labour market region, 32 percent moved to a locality within the same state and 12 percent moved interstate. Most of those aged 55-69 who moved labour market region, did so away from large urban centres (ABS 2004). The key sending localities were Sydney, Melbourne, Brisbane, Adelaide, Perth and the ACT. Popular destinations for retirees included the coastal towns of central and south Queensland, of New South Wales, the Gippsland region in Victoria, the hill and coastal regions surrounding Adelaide in South Australia and the coastal towns south of Perth in Western Australia. In short, non metropolitan, coastal localities within 300 kilometres of capital cities are the most popular destinations for Australian retirees (ABS 2004, Sander 2010). Sander's (2010) research, together with that of Bohnet and Moore (2011), Connell and McManus (2011) and Ragusa (2011) suggest that Australian retirees are however now more likely than ever to move to small inland rural towns. The emergence of elderly migration flows from metropolitan regions to small inland rural towns could signify the beginning of a much broader socio-economic shift occurring in parts of rural Australia.

To understand more about the growth and spatial and temporal unevenness of exurban elderly migration flows to small towns in rural Australia, the remainder of this section fleshes out three core concepts of elderly migration. These are that individuals' propensity to migrate and selection of destination changes over the life course; place based push and pull factors of migration underpin migrants' decisions to migrate and also selection of destination; and, migration flows to places increase in momentum over time as the first migrants establish social networks, infrastructure and economic activities, which, in turn, help to ease the way for future in-migrants to settle.

Life Course Migration

Migration decisions are linked to phases in the life course and, as such, the migrations of the elderly have been considered separately to those of working age and young people (Litwak and Longino 1987, Longino and Haas 1992, Stockdale *et al.* 2013). In Australia, as in many other developed countries, migration often occurs as a response to life course events (Curry *et al.* 2001, Davies 2008, Gibson and Argent 2008, Stimson 1998, Walmsley *et al.* 1998). The first move people usually make away from the family home is to access employment and/or education. Later in life, as people approach retirement, they may move in response to a decrease in their income, to be closer to family, to access an environment that offers improved lifestyle opportunities or a combination of these (Hugo and Smailes 1985, Stockdale 2006).

While Australian research on life course migration specifically focused on elderly populations is limited, there is a developed international literature that provides useful insights for Australia. International studies on elderly migration have identified three groups of elderly migrants (see Litwak and Longino 1987); these include those who make a move immediately after retirement, largely for amenity reasons but also, in some instances, to be nearer to family and social networks; those who make a move to be near to a primary caretaker, such as a family member, following the onset of moderate disability, and; those who make a move into an institutional setting where disability is such that family members or friends are not able to take on the care responsibility. However this classification approach has been criticised for overlooking important variations in elderly migration patterns. For example, Lovegreen *et al.* (2010) examined elderly people's moves after initial retirement migration. They found that elderly amenity migrants, post migration, could be part of one or more of the following migration types: moving to access new amenity and lifestyle opportunities with little consideration to future care needs; moving to an independent living section of a continuing care retirement community; moving to live with, or near, family to access informal assistance; moving to an assisted living facility; or moving into a nursing home.

Associated with life course migration are distinct geographical patterns. For instance, those moving into assisted living or nursing homes often locate in large urban centres where such facilities are readily available. Contrastingly, older people moving for amenity reasons often seek small towns at the periphery of large metropolitan regions. The geographical nature of elderly migration trends associated with phase of the life course has specific implications for understanding elderly migration in rural Australia, although since many small towns in rural Australia have limited healthcare and assisted living facilities it is unlikely that people in their late old age would move there. However, many small rural towns have desirable amenity characteristics, making them 'hot spots' for in-migration of elderly people during their early old age. This dissonance will be discussed further later in the chapter.

At this point, however, it is useful to consider how place-based push and pull factors of migration influence elderly migration trends. The following section reviews the core tenets of what is now a well developed body of literature on the relationship between the geographic characteristics of places and elderly migration trends.

Place Based Push and Pull Factors of Migration

Migration is influenced by place based factors at the origin and destination. Commonly termed 'push and pull factors', place based factors such as the condition and character of the physical and cultural environments and relative access to services and leisure opportunities are all central to migrants' selection of destination. For those moving to small town rural Australia, the amenity of the natural and built environments has been identified as critically important to migrants' decisions to move and selection of destination (Hugo and Smailes 1985, Walmsley *et al.* 1998), while close-knit communities which provide personal interaction contribute to stability (Winterton and Warburton 2012).

Amenity migration to rural Australia, often termed 'sea change' or 'tree change' migration (terms which reflect migrants' preferences for coastal or forested regions) is one of the most significant counterurbanisation migration flows in Australia. Across rural Australia, areas with certain desirable environmental and other geographic qualities have long experienced high rates of in-migration (Argent *at al.* 2007). However, despite the attention that amenity migration flows to rural Australia have received in recent years, studies have rarely separated out elderly migrants from other age groups. Australian research on amenity migration has also remained largely silent on the process of amenity migration. In particular, it is not well understood whether people select a destination to move to before making the decision to move, or if people decide to move and then select a destination. Understanding how people come to make their decision to migrate and the role that place plays in this process is crucial to efforts to plan for the spatial distribution of public resources for an ageing population.

To address this gap in knowledge about Australian elderly amenity migration, insights can be gained from international literature. Of most relevance to understanding elderly amenity migration in Australia is the work on the amenity migration process for elderly populations by Haas and Serow (1993), who explored the importance of place based factors to elderly migration decisions, proposing a model for the amenity retirement migration process. They found that amenity retirement migrations followed a period of consideration, where the focus was often on the undesirable (or push) factors at the point of origin and attractive factors at the destination. Haas and Serow did not prescribe a time frame for this period of consideration, but noted that migration usually followed a phase of 'remote thoughts' about a move and then a period of more serious investigation and decision-making. About half of all elderly migrants only considered moving to one single destination prior to their migration, and these destination-specific migrants were usually familiar with their destination from previous work or holiday visits.

While there has not any been any comparable comprehensive study of Australian elderly migration flows, Davies and James (2011) considered the movement of people aged 60+, most of who were retired, from the metropolitan area of Adelaide (capital city of South Australia) to Victor Harbor. Victor Harbor is located one hour's drive south of Adelaide and has long been a coastal retirement destination. The high amenity value of the natural environment is considered the major factor underpinning the retirement migration flow. Of those who had relocated from Adelaide most had only considered moving to Victor Harbor. Likewise, those who moved to Victor Harbor from other rural areas in South Australia were also 'destination specific', with Victor Harbor the only location they seriously considered moving to prior to their decision to migrate. The ability to purchase or build a house that would be suitable to their needs in late old age, the presence of healthcare services that supported ageing-in-place, and the availability and quality of leisure facilities and opportunities were identified as the most important factors underpinning the decision to move to Victor Harbor. One resident commented:

> I wanted to go somewhere I know there is good old age facilities and I didn't want to be put there when I'm 70 or 75 and I'm totally new to the area. [Moving

here] is a view to the future, that was one of the things we were looking at… so
we were looking where we can settle in for the rest of our life if necessary and
get to know the area and get to know the shopping centres and get to know the
people from the church and get to know just a few people before they put us into
a nursing home (Davies and James 2011: 136).

Another resident commented 'The area is much fresher, it's a much healthier
place to live, a number of good walks. We are both very active and still engage in
water sports, tennis, golf and walking' (Davies and James 2011: 136). Place based
factors influenced both the decision to migrate and choice of destination.

Victor Harbor demonstrated that retirees viewed it as an established retirement
destination. This 'established' nature of Victor Harbor was an important factor in
influencing retirees' decisions to move to the town. It was widely perceived that
Victor Harbor provided many of the specific services and infrastructure required
by elderly people, particularly as they progressed into late old age. This temporal
nature of the elderly migration flow to Victor Harbor is reflective of the broader
trend whereby elderly in-migration flows build over time.

The Temporal Nature of Migration Flows

Research on elderly migration has revealed the importance of the non-static nature
of the push and pull factors operating at the donor and destination localities to
migration flows. The factors that compel an individual or couple to move at one
point in time may change considerably over the years following their migration.
For example, amenity migrants usually select their destination based on the
quality and quantity of amenities available. In some locations, this creates a higher
demand for amenities, and in some cases reduces access and/or reduces the quality
of the amenities (Ragusa 2011). For some migrants this reduced access or quality
might propel them to consider moving to another location (Ragusa 2010, 2011).

Migration into rural areas becomes less uncertain over time as the first migrants
establish social networks, infrastructure and economic activities, which help to
ease the way for future in-migrants to settle (Tonts and Greive 2002). Indeed, urban
to rural migration, particularly at or during retirement, has been 'normalised' in
Australian society over the last decade as evidenced by the existence of more than
twenty businesses specialising in assisting people to make an urban-rural move. A
rapid assessment of 'blog' pages also reveals a large informal support network for
people looking to move from urban areas into rural towns. Perhaps most tellingly,
there has also been a reality television series developed called 'The Real Sea-
Change' which documented the experiences of families, couples and individuals
as they relocated from large cities to rural towns.

As urban to rural elderly migration has become increasingly normal over time,
a new trend has emerged where older Australians seek out non-traditional rural
retirement destinations – towns that are off the beaten track. While elderly in-
migration to traditional migration destinations (coastal areas within 300 kilometres

of capital cities) remains strong, for a growing number of retirees the smaller, inland, usually less well-serviced and typically un-gentrified rural towns are sought after. Sander (2010) found that from the early 2000s a number of established non-metropolitan coastal retirement destinations were experiencing a decline in net gains of retirees, due to both an overall decrease in in-migration and an increase in out-migration. By contrast, areas that were previously peripheral retirement destinations had experienced a growth in in-migration. A number of inland rural areas have emerged as popular retirement migration destinations including the Darling Downs in south-east Queensland, Murray in NSW, the northern region of Tasmania and the lower southern regions in Western Australia.

Migration patterns across rural Australia closely reflect the social and economic functions and fortunes of rural places. Migration flows are both a function of and integral to rural change. As a community develops new social and economic activities and new residents are attracted, they bring with them resources and establish new activity patterns which, in turn, drive further social and economic change (Tonts and Greive 2002). The association of elderly migration flows with both life course phase and geographical factors means that elderly migration flows are not uniform across rural Australia, nor are they uniform over time. This diversity is crucial to efforts to plan for the distribution, renewal and conservation of resources in both donor and recipient communities. The following section considers how urban to rural elderly migration is shaping new socio-economic functions and driving renewal in some parts of rural Australia.

Elderly In-Migration and the Production of New Socio-Economic Spaces in Small Town Rural Australia

As outlined above, the literature on why elderly Australians move from urban to rural settings can be described as patchy. There is no one reason why elderly people move. For some, their move is underpinned by a desire to participate in new social and leisure opportunities. For others, the ability to secure affordable housing is the core factor underpinning their move. Others seek out a rural setting for their retirement in an effort to reconnect with a past way of life or, if they previously lived in the town to which they move, to reconnect with family and social networks. While reasons why people choose to move and select destinations are complex it is clear that the decision remains a highly personal one. Individuals consider their own resources (including personal history), their desires and perceptions about the opportunities that another place can afford them in both their decision to move and selection of a destination (Davies and James 2011).

The following discussion on why retirees are moving to small towns in rural Australia, and how this migration flow is reshaping the way rural spaces are consumed and produced, draws on the recent history of the New South Wales town of Gloucester. Gloucester, a small inland rural town, is located roughly 260 kilometres from the State capital of Sydney and 70 kilometres from the coast. It is situated in a region of high natural amenity value, with the town located on the Gloucester River and shadowed by a dramatic world heritage listed mountain range known as 'Barrington Tops'. Gloucester is not a traditional amenity or counterurbanisation destination. It is not located on a major transport route and had an established identity as an agricultural service centre with the town's economic functions closely linked to the dairying, cattle production and timber activities of its surrounding agricultural hinterland. Interviews were conducted with people involved in community endogenous development efforts, including those who had lived in the town their whole lives, or a good part thereof, as well as people who had moved to the town within the last five years (Davies 2005).

Gloucester was established in 1855 by the Australian Agricultural Company and developed during the 1900s to service the forestry and dairying industries of the region. As these industries grew so too did Gloucester's population and the scope of its social and economic activities. Gloucester was home to a vibrant dairy processing centre, timber mills, a number of local schools, a well-staffed hospital, and a variety of small businesses. It also had established social clubs, and 'traditions' such as the annual Royal Agricultural Show. During the mid 1990s the broader region suffered considerable job losses, resulting from reforms in both forestry and dairying. However, since the mid 1990s, Gloucester has experienced a considerable shift in its functions and fortunes. These reforms and job losses had a flow on impact for the town, with the dairy factory, timber mills and storage facilities closing. Businesses directly supporting those activities, such as transport and warehousing, also downsized or closed. Consequently, businesses that were not directly affected by the initial closures experienced a period of economic hardship with fewer customers and limited opportunities for expansion. One resident commented:

> Gloucester appeared to be dying. If you went for a drive down the main street you would see lots of vacant stores. There was little prospect for jobs for young and others who were unemployed and so young people were leaving the town in droves. Who can really blame them when even the long time residents were thinking that the place was on its last legs. People didn't want to renovate their homes or invest in businesses as they saw there would be little return as the town was going to keep declining.

Job losses and limited potential for growth resulted in increased outmigration and a sharp decline in in-migration, with a net negative migration rate. The local property market, both commercial and residential, experienced a downturn. Two of the dominant features of the property market at that time were the number of low-rent commercial vacancies in the town and the increase in the number of small 'hobby' farms that were available due to an increase in the number of farmers subdividing and selling their land.

The decline of the traditional functions of the town proved necessary to enable the conditions to emerge that would attract the 'first wave' of ex-urban retirement migrants. This 'first wave' of retirement migrants moved to Gloucester during the late 1990s; most were retired or semi-retired and independently financially secure. They were attracted to the region due to its high natural amenity value and distance from major centres, and specifically to Gloucester due to the relatively low cost of property. In particular, the relatively low cost small 'hobby farms' surrounding the town drew retirees and semi-retirees to the town. While undoubtedly the high natural amenity of the region played a significant role in attracting in-migrants to the region, the cost of housing was influential in choosing what town in which to settle within the region. One retirement migrant who moved to Gloucester during the late 1990s commented:

> Unlike the people who have spent their whole lives here, we have travelled the world living in many different places. We chose Gloucester to live in. We are not here by circumstance, we have picked this place as the community that we want to live in. So we are keen to see the town go well. We saw the potential when we moved here.

In response to the closure of the dairy factory, closure of timber mills and other associated job losses, the Gloucester town council upgraded the main street in an attempt to stimulate new business activity. This attracted federal government funding and involved resurfacing the road, adding road furniture to slow traffic, creating road side parking, adding trees and garden beds, repaving the footpaths and adding street art. Unemployed workers from the dairy factory were offered jobs within the project, and community members were encouraged to become involved. A range of other community initiatives to rebuild community spirit and encourage economic investment in the town were also developed. Some were lead by the council or established community organisations and others were more informal groups of interested residents. This activity and residents' willingness to invest time, effort and often money into the town proved particularly important in attracting the first wave of elderly in-migrants.

Many in-migrants commented that the Gloucester council's effort to redevelop the main street demonstrated a strong sense of community and commitment to place. The initiative of the council to employ those workers who had lost their jobs in the dairy factory on the main street upgrade project demonstrated that the community were 'willing to look after their own'. One resident commented 'The Council showed people that the town still did have a future if they were willing to work for it and create it'. Community initiatives to restructure the town's economic activities generated an openness within the business community to new investment and new ideas and within local social networks to new people, which attracted in-migrants, many of whom sought to be involved in community development activities.

Following initial community led efforts to turn the fortunes of the town, Gloucester began experiencing considerable economic investment. Tourism quickly developed from being a very marginal activity with few employees and limited infrastructure into one of the most important sectors in the town, with an annual turnover of more than $20 million and directly employing 7 percent of the local workforce (ABS 2007, Hunter Development Brokerage 2006). The development of tourism also strengthened the retail sector, which provided employment for 11 percent of the local workforce (ABS 2007). One resident commented 'The future is looking brighter for Gloucester Shire. As confidence grows among the people, we are sure that more innovative ideas will come forth in tourism and other business sectors'. The expansion of tourism activities in Gloucester was greatly enhanced by the activities and investments of in-migrants. In particular, a few retired in-migrants became involved in community initiatives to develop festivals and events to be held in Gloucester. These festivals and events, including Snowfest, generated an important market for local tourism and accommodation providers and stimulated further innovation and investment (see Davies 2011).

As Gloucester shifted away from being a declining agricultural services centre and generated new social and economic activities (largely based around tourism), the town quickly gained a reputation as being an 'off the beaten track' retirement destination (Gloucester Shire Council 2009). In particular, Gloucester became an attractive destination for those who were looking to move away from urban areas but not to traditional coastal and rural retirement destinations. In this sense, the activities of those migrants who had moved to Gloucester in the mid-to-late-1990s, helped to generate the conditions necessary to promote and support a larger in-flow of elderly (both retired and non-retired) into the town.

Gloucester is a vibrant, multifunctional town. Although it could not be considered a major 'tree change' destination, it has a growing flow of elderly in-migrants. Over the last 15 years the activities of elderly in-migrants have helped to reshape its socio-economic characteristics. Now, as the initial in-migrants are moving into late old age, they are again stimulating a shift in the functions of the town, through an increase in demand for health care and social wellbeing support – both in-home and residential.

The Gloucester case study illustrates how elderly urban to rural migration flows are influenced by the social, economic and geographical factors of the destination community, and how elderly migration flows build over time and can influence the socio-economic functions and fortunes of rural towns over the long term. While the exact conditions and circumstances that underpinned Gloucester's initial wave of elderly migration, and then supported the growth of this migration, will not occur elsewhere, the general tenets of Gloucester's recent development history are unremarkable. Indeed, many small rural towns across Australia have undergone renewal and reinvention, and have attracted new flows of in-migrants which, in various ways, are shaping new futures for Australia's small rural towns.

Conclusion

Individuals' decisions to move (or not) and selection of destination shift over the life course. As people move through different phases of the life course, they react differently to the social, economic and environmental characteristics and resources of places. For elderly people there are five broad categories of moves. The type of move someone will make is determined by their personal social, economic and demographic history and access to social and economic resources. In general, those moving from urban to rural areas are 'amenity movers' and tend to be in their early old age.

Over the last two decades, rural Australia has experienced a steady increase in the number of elderly people moving from urban areas into rural towns. The vast majority of these moves have been to relatively nearby coastal destinations. However, during the first decade of the 2000s there has been an increase in the number of older people seeking out more remote inland towns – retirement destinations that are 'off the beaten track'. Not all inland rural towns experience elderly in-migration now or in the future. Those towns high in natural amenity value, with an affordable stock of quality housing and an active, engaged community are the most likely to attract elderly in-migrants. The implications of elderly in-migration for small rural communities include increased demand for health care and social welfare infrastructure, increased incidence of social isolation as people enter late old age, increased pressure on housing and, on a more positive note, increased rates of participation in social groups and community organisations. As Australia's aged population grows and more people choose to move to rural towns for their retirement the need to understand the processes and implications of urban to rural migration will become more urgent.

References

Argent, N., Smailes, P. and Griffin, T., 2007, 'The Amenity Complex: Towards a Framework for Analysing and Predicting the Emergence of a Multifunctional Countryside in Australia', *Geographical Research*, 45: 217-232.

Australian Bureau of Statistics, 2004, *Australian Social Trends*, Canberra: Australian Bureau of Statistics.

Australian Bureau of Statistics, 2007, *Population and Housing Census Time Series Profiles (selected tables)* [Online]. Canberra: Australian Bureau of Statistics. Available at: http://www.censusdata.abs.gov.au.

Bohnet, I.C. and Moore, N., 2011, 'Sea- and Tree-Change Phenomena in Far North Queensland, Australia: Impacts of Land Use Change and Mitigation Potential', in G. Luck, R. Black and D. Race (eds), *Demographic Change in Australia's Rural Landscapes: Implications for Society and the Environment*, Dordrecht: Springer, 45-70.

Connell, J. and McManus, P., 2011, *Rural Revival? Place Marketing, Tree Change and Regional Migration in Australia*, Aldershot: Ashgate Publishing.

Costello, L., 2007, 'Going Bush: The Implications of Urban-rural Migration', *Geographical Research*, 45: 85-94.

Costello, L., 2009, 'Urban-rural migration: Housing Availability and Affordability', *Australian Geographer*, 40: 219-233.

Curry, G.N., Koczberski, G. and Selwood, J., 2001, 'Cashing Out, Cashing In: Rural Change on the South Coast of Western Australia', *Australian Geographer*, 32: 109-124.

Davies, A., 2005, 'The Role and Nature of Local Leadership in Influencing the Socio-Economic Viability of Small Rural Towns in Australia' (unpublished PhD thesis), Armidale: University of New England.

Davies, A., 2008, 'Declining Youth In-migration in Rural Western Australia: The Role of Perceptions of Rural Employment and Lifestyle Opportunities', *Geographical Research*, 46: 162-171.

Davies, A., 2009, 'Understanding Local Leadership in Building the Capacity of Rural Communities', *Geographical Research*, 47: 380-389.

Davies, A., 2011, 'Local Leadership, Rural Revitalisation and Festival Fun', in C. Gibson and J. Connell (eds), *Festival Places: Revitalising Rural Australia*, Bristol: Channel View, 61-73.

Davies, A. and James, A., 2011, *Geographies of Ageing: Social Processes and the Spatial Unevenness of Population Ageing*, Aldershot: Ashgate Publishing.

Gibson, C. and Argent, N., 2008, 'Getting On, Getting Up and Getting Out? Broadening Perspectives on Rural Youth Migration', *Geographical Research*, 46: 135-138.

Gurran, N., 2008, 'The Turning Tide: Amenity Migration in Coastal Australia', *International Planning Studies*, 13: 391-414.

Haas, W.H., and Serow, W.J., 1993, 'Amenity Retirement Migration Process: A Model and Preliminary Evidence', *The Gerontologist*, 33: 212-220.

Hugo, G.J. and Smailes, P.J., 1985, 'Urban-rural Migration in Australia: A Process View of the Turnaround', *Journal of Rural Studies*, 1: 11-30.

Hunter Development Brokerage, 2006, *Gloucester Local Environmental Study*, Gloucester: Gloucester Shire.

Kinsella, K. and He, W., 2009, *An Ageing World: 2008, U.S. Census Bureau International Population Reports P95/09-1*, Washington DC: United States Government Printing Office.

KPMG, 2012, *Food Security and Asia-Australia Relations. Australia in the Asian Century: Executive Summary*. KPMG. Available at: http://www.kpmg.com/AU/en/IssuesAndInsights/ArticlesPublications/Documents/food-security-and-asia-australia-relations-exec-summary.pdf.

Litwak, E. and Longino, C.F., 1987, 'Migration Patterns among the Elderly: A Developmental Perspective', *The Gerontologist*, 27: 266-272.

Longino, C.F. and Haas, W.H., 1992, 'Migration and the Rural Elderly', in C. Bull (ed.), *Aging in Rural America*, London: Sage, 17-29.

Lovegreen, L.D., Kahana, E. and Kahana, B., 2010, 'Residential Relocation of Amenity Migrants to Florida: "Unpacking" Post-amenity Moves', *Journal of Aging and Health*, 22: 1001-1028.

National Rural Health Alliance, 2009, *Ageing in Rural, Regional and Remote Australia*, Canberra: National Rural Health Alliance.

National Seniors Australia Productive Ageing Centre, 2010, *Getting Involved in the Country: Productive Ageing in Different Types of Rural Communities*, Canberra: Government of Australia, Department of Health and Ageing.

Productivity Commission, 2005, *Economic Implications of an Ageing Australia: Productivity Commission Research Report*, Melbourne: Commonwealth of Australia.

Ragusa, A.T., 2010, 'Country Landscapes, Private Dreams? Tree Change and the Dissolution of Rural Australia', *Rural Society*, 20: 137-150.

Ragusa, A.T., 2011, 'Seeking Trees or Escaping Traffic? Socio-cultural Factors and "Tree-change" Migration in Australia', in G. Luck, D. Race and R. Black (eds), *Demographic Change in Australia's Rural Landscapes: Implications for Society and the Environment*, Dordrecht: Springer, 71-100.

Sander, N.D., 2010, 'Retirement Migration of the Baby Boomers in Australia: Beach, Bush or Busted?' (unpublished PhD Thesis), Brisbane: The University of Queensland.

Spies-Butcher, B., 2011, 'The Myth of the Ageing "crisis"', *The Conversation*, 27 April 2011. Available at: http://theconversation.edu.au/the-myth-of-the-ageing-crisis-168.

Stimson, R.J., 1998, 'Why People Move to the "Sun-belt": A Case Study of Long-distance Migration to the Gold Coast, Australia", *Urban Studies*, 35: 193-214.

Stockdale, A., 2006, 'The Role of a Retirement Transition in Repopulation of Rural Areas', *Population, Space and Place*, 12: 1-13.

Stockdale, A., MacLeod, M. and Philip, L., 2013, 'Connected Life Courses: Influences On and Experiences of "Midlife" In-migration to Rural Areas', *Population, Space and Place*, 19: 239-257.

Tonts, M. and Greive, S., 2002, 'Commodification and the Creative Destruction in the Australian Rural Landscape: The Case of Bridgetown, Western Australia', *Australian Geographical Studies*, 40: 58-70.

Walmsley, D.J., Epps, W.R. and Duncan, C.J., 1998, 'Migration to the New South Wales North Coast 1986-1991: Lifestyle Motivated Counterurbanisation', *Geoforum*, 29: 105-118.

Winterton, R. and Warburton, J., 2012, 'Ageing in the Bush: The Role of Rural Places in Maintaining Identity for Long Term Rural Residents and Retirement Migrants in North-east Victoria, Australia', *Journal of Rural Studies*, 28: 329-337.

Chapter 4

Immigrant Settlement in Regional Australia: Patterns and Processes

Graeme Hugo

Introduction

Few countries have been more influenced by international migration in the contemporary era than Australia. In 2011, 27 percent of the national population were foreign-born. A further 20 percent were second generation immigrants and at any one time more than a million foreigners are temporarily in the country. In the last decade, however, while most immigrants continue to settle in the largest cities, there has been an increase in the numbers settling elsewhere. The numbers of overseas-born persons living outside these 'gateway cities' increased by a third from 771,574 in 2001 to 1,001,645 in 2011. This represents a small but significant shift, with a reversal of longstanding trends of substantially greater growth in the capital cities.

Similar changes have occurred in other high-income immigrant destination countries. In the United States there has been a significant decentralisation of immigrant settlement: 'Immigrants now settle in small towns as well as large cities and in the interior as well as the coasts' (Hirschman and Massey 2008: 3). Much the same has occurred in mainland Europe, notably in Spain (Oliva 2010), the United Kingdom (Green *et al.* 2012) and New Zealand (Spoonley and Bedford 2012). A distinctive part of the Australian experience, however, has been explicit policy intervention to facilitate immigrant settlement beyond the major cities.

This chapter begins with an examination of the pattern of immigrant settlement in Australia, especially in non-metropolitan areas, and how and why this has changed over time. Contemporary patterns differ somewhat from traditional settlement of migrants outside cities. Demographic and economic changes are important drivers but policy has also played a role. Anticipated demographic and economic trends suggest that international migration will play a bigger role in economic and population growth in non-metropolitan areas in the future and that immigrants will be an increasingly important part of regional development in Australia.

Patterns of Immigrant Settlement in Regional Australia

Discussions of the dynamics of changing population distribution in Australia usually focus exclusively on issues of internal migration but where migrants choose to settle (and not settle) is an increasingly important influence on national and regional population growth rates and helps shape changes in population distribution. The spatial distribution of immigrants is never identical to that of the host population. In 2011, 81 percent of the overseas-born population lived within the boundaries of the Greater Capital City areas compared with only 59 percent of the Australia-born. However, this concentration was not always the case. Migrant groups have played an important role in the development of several parts of regional Australia over the last 150 years. Germans, for example, were heavily involved in agricultural expansion in several colonies in the nineteenth century while Italians were significant in development of the sugar cane, market gardening and intensive horticultural industries in the first half of the twentieth century (Borrie 1954). Substantial numbers of Germans lived outside Adelaide in 1891 while many Italians lived outside Brisbane in 1933 (Borrie, 1954: 179).

Prior to World War II, immigration to Australia was overwhelmingly of people of British origin, although small numbers of people of more diverse backgrounds played a significant role in regional development. The Chinese influx associated with the gold rushes of the nineteenth century saw the growth of significant communities in regional centres of the eastern states, although they contracted in size after the introduction of the White Australia policy in 1901 (Choi 1974). Southern European settlement in the pre-war period was limited but spatially concentrated. Price (1963: 11) estimated that the Southern Europe-born population increased from around 6,000 at the 1891 census to 60,450 in 1947. In 1947, if the second generation are included, Southern Europeans numbered almost 100,000 or 1.3 percent of the national population and more than a half of them lived outside large metropolitan areas (Table 4.1).

Moreover many lived in rural areas with only 9 percent being in provincial cities, although for Greeks it was close to a fifth. Southern European engagement in intensive

Table 4.1   Main Southern European Birthplace Groups: Distribution, 1947

Birthplace	n	Percent			
		Migratory*	Rural	Provincial City	Metropolitan
Italy	33,700	0.2	51.6	5.2	43.0
Greece	12,500	0.5	24.3	18.2	57.0
Malta	3,300	1.2	31.4	5.4	62.0
Yugoslavia	8,050	0.3	54.5	7.2	38.0
Total	**57,550**	**0.3**	**44.2**	**8.5**	**47.0**

* In early postwar censuses a category of 'migratory' was identified in the census for people who were itinerant and had no fixed place of residence.

Source: Price, 1963: 158

horticulture, viticulture and irrigated agriculture saw them heavily concentrated in market gardening areas close to cities, irrigated areas along major rivers, coastal regions of sugar cane and fruit growing and, to a lesser extent, in fishing and mining localities. They avoided the extensive wheat-sheep belt and remote pastoral areas where the population was dominated by British-origin Australians and, to a lesser extent, by the indigenous population.

Following World War II, Australian immigration underwent several transformations of unprecedented scale and increasing diversity (Figure 4.1). Previously, immigration was dominated by British settlers but following the war there were successive waves from Eastern Europe, Southern Europe, the Middle East and – following the dismantling of the White Australia policy in the 1970s – from Asia and Sub-Saharan Africa. Post-war immigrant settlement was also different in that it was overwhelmingly focused on large metropolitan centres, following rapid industrialisation in the early post-war decades; increasing urbanisation between 1947 and 2001 saw a significantly greater concentration in capital cities among the overseas-born than the Australia-born (Table 4.2).

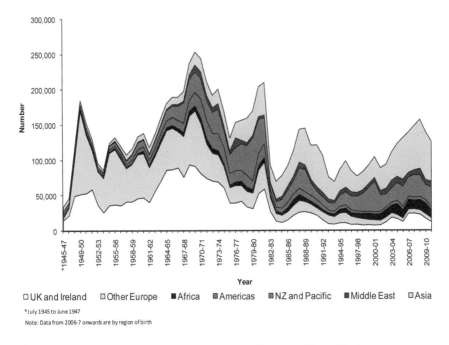

Figure 4.1   Australia: Settler Arrivals by Region of Last Residence, 1947-2011

Source: DIMIA, Australian Immigration Consolidated Statistics; DIAC, Immigration Update, various issues; and DIAC unpublished data

Table 4.2   Distribution of Overseas-Born Population Between Capital Cities and Rest of State, 1947, 2001, 2006 and 2011

	1947		2001		2006		2011		Growth Rates		
	Number	%	Number	%	Number	%	Number	%	1947–2001	2001–06	2006–11
Major Capital Cities	453368	61.8	3307577	81.1	3557486	80.6	4278495	81.0	3.75	1.47	3.76
Rest of States	280004	38.2	771574	38.2	857873	18.9	1001645	19.0	1.89	2.14	3.15
Total	733372	100.0	4079151	100.0	4415359	100.0	5280140	100.0	3.23	1.60	3.64

Source: ABS Censuses

Between 2001-06 a small but significant change occurred. For the first time since World War II there was a faster increase in the overseas-born living outside the capital cities than within them, resulting in a small increase in the proportion living there. In 2006-11 there was a return to a slightly faster rate of growth in the capital cities and hence a small reduction in the proportion living outside the capitals. Nevertheless rapid growth occurred in both the capitals and non-metropolitan areas during this period of unprecedentedly high national immigration (Table 4.2). This increasing urbanisation of the overseas-born has been particularly marked for recently arrived immigrants, and for people from culturally and linguistically diverse (CALD) backgrounds (Table 4.3). The proportion of new arrivals settling in capitals was 89 percent for those arriving in 1981-86 and 90 percent for 1996-2001. The same pattern is present but less marked among those from mainly English-speaking (MES) countries with 77 percent and 70 percent respectively.

Table 4.3   Number and Percentage of Overseas-Born Persons Resident in Capital Cities by Origin and Length of Residence, 1986, 2001, 2006 and 2011

	0-4 Years		5+ Years	
	Number	*Percent*	*Number*	*Percent*
1986				
MES Origin	142,722	76.9	890,809	73.2
LOTE Origin	240,864	88.6	1,245,254	83.8
Total Overseas-Born	383,586	83.9	2,136,063	79.0
2001				
MES Origin	145,936	77.0	936,796	70.2
LOTE Origin	307,781	90.1	1,762,488	86.2
Total Overseas-Born	453,717	85.4	2,699,284	79.9
2006				
MES Origin	173,293	74.2	943,568	69.4
LOTE Origin	416,389	88.8	1,857,957	86.8
Total Overseas-Born	589,682	83.9	2,801,525	80.0
2011				
MES Origin	225,518	74.3	1,047,149	69.2
LOTE Origin	617,114	87.7	2,205,445	87.4
Total Overseas-Born	842,633	83.7	3,252,594	80.6

Source: ABS Censuses

Settlement of immigrants in non-metropolitan areas in the second half of the twentieth century tended to follow pre-war spatial patterns. Migrants particularly concentrated in industrialising regional cities, like Newcastle, Wollongong, Geelong and Whyalla; CALD groups also continued to concentrate in intensive agricultural areas and eschew the wheat-sheep and pastoral areas. The Southern Europe-born population of one community on the River Murray in South Australia, for example, increased from 90 to 501 between 1947 and 1971 (Hugo 1975). This included both newly arrived immigrants and others who had spent some time in capital cities. Chain migration processes operated with most settlers coming from a small number of districts in Greece, Italy and elsewhere; many had previously lived in rural areas and had been farmers in their homeland.

At the 2006 census, 63 percent of the Australia-born lived in major cities compared with 93 percent of the CALD and 76 percent of the MES-born population. For recent arrivals of CALD and MES migrants the proportions are 92 percent and 83 percent respectively. There is some evidence of a slight lessening of the dominance of the capital cities in the initial settlement of migrants, reflecting a continuing pattern of new immigrants, especially those from CALD origins (more than two thirds), settling outside the capitals. Immigrants from MES countries, especially New Zealand and the United Kingdom, although more concentrated in

Table 4.4   Australia: People Born Overseas by Remoteness Area, 2001 and 2006

Remoteness Area	Overseas-Born 2001 ('000)	Overseas-Born 2006 ('000)	% Overseas-Born 2001	% Overseas-Born 2006	Growth Rate 2001-06
Major Cities	3409.0	4825.6	83.0	80.6	7.20
Inner Regional	431.7	668	10.5	11.2	9.12
Outer Regional	208.9	378.8	5.1	6.3	12.64
Remote	35.0	75.1	0.9	1.3	16.50
Very Remote	19.5	36.1	0.5	0.6	13.11
Total	**4105.6**	**5983.6**	**100.0**	**100.0**	**7.82**

Note: *Highly Accessible Major Cities* – Locations with relatively unrestricted accessibility to a wide range of goods and services and opportunities for social interaction. *Accessible Inner Regional Areas* – Locations with some restrictions to accessibility of some goods, services and opportunities for social interaction. *Moderately Accessible Outer Regional Areas* – Locations with significantly restricted accessibility of goods, services and opportunities for social interaction. *Remote Areas* – Locations with very restricted accessibility of goods, services and opportunities for social interaction. *Very Remote Areas* – Locationally disadvantaged – very little accessibility of goods, services and opportunities for social interaction (ABS)

Source: ABS 2001 and 2006 Censuses

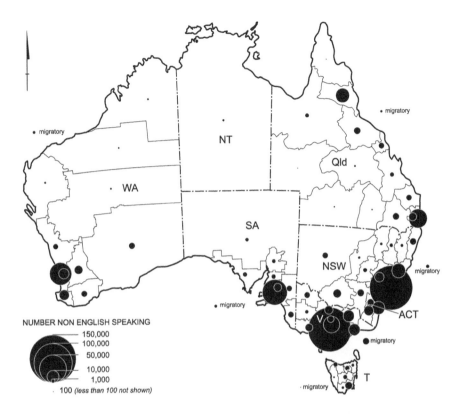

Figure 4.2   Distribution of Non-English-Speaking Persons by Statistical Division, 1954

Source: http://www.abs.gov.au/AUSSTATS/abs@.nsf/DetailsPage/2108.01954?Open Document

major cities compared with the Australia-born, are more similar to them than those from CALD origin countries. For both groups, especially the MES group, there is a strong tendency with increasing length of residence in Australia for settlement patterns to converge toward those of the Australia-born (Hugo and Bell 2000). Another dimension of change is evident in the pattern of immigrant settlement according to degree of accessibility; the growth rate of the overseas-born has recently been greatest in remote areas mainly due to the impact of the mining industry (Table 4.4).

The distribution of the CALD-born population in 1954 (Figure 4.2) demonstrates that the largest numbers were in the state capitals, especially Sydney and Melbourne, but there was a strong representation in non-metropolitan areas, especially in coastal areas and around the cities. By 2011 there were clear differences (Figure 4.3). The dominance of the capitals, especially Sydney and Melbourne, was more apparent but there were also significant numbers in non-metropolitan areas. Over the 2009-12 period some 15 percent of all permanent

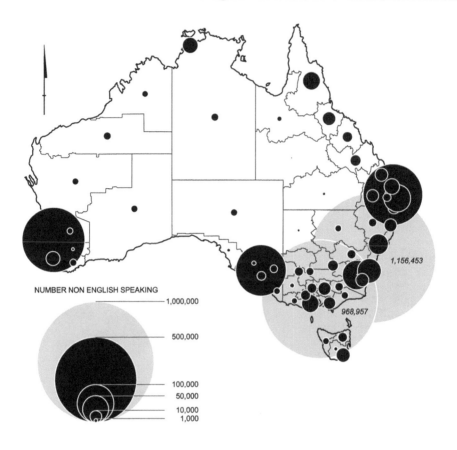

Figure 4.3   Distribution of the CALD-Born Population by SA4 Region, 2011

Source: ABS 2011 Census

arrivals gave non-metropolitan destinations on their arrival in Australia. The places nominated by 2009-12 arrivals as their destination (Figure 4.4) emphasise that a substantial number nominated non-metropolitan locations.

A key feature of immigrant settlement, especially of those from CALD backgrounds, is a high degree of spatial concentration. This varies between different origin groups and can be measured using the Index of Dissimilarity (the I_D) (This index is the percentage of a particular subpopulation which would have to change their place of residence if the distribution of that group between sub-areas of the region under study is to become exactly the same as that of the other subgroup. An index of 0 would mean that the two subpopulations had exactly the same relative distribution while an index value of 100 represents a complete

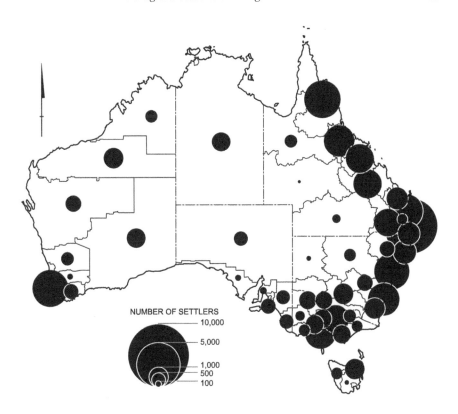

NUMBER OF SETTLERS
— 10,000
— 5,000
— 1,000
— 500
— 100

**Figure 4.4   Settlers Arriving between July 2009 and June 2012
by Statistical Divisions**

Source: DIAC Settlement Reporting Facility

'apartheid' situation, with no person of one sub-group living in the same sub-area as people of the other subgroup. These two extremes rarely occur. If the index is less than 20 there is little spatial separation of the two sub-populations, if it exceeds 30 there is some significant separation and if it exceeds 50 there is very significant separation).

Wide'differences exist between birthplace groups in their propensity for concentration (Table 4.5). The highest I_Ds tend to be for those groups that have come to Australia as refugee-humanitarian settlers, such as those from Iraq (65.8) Bosnia and Herzegovina (66.8), Former Yugoslavia (79.2) and Turkey (66.2). More recently arrived groups like those from Afghanistan and Sub-Saharan African nations also have high levels of concentration, while

Table 4.5   Australia: Non-Metropolitan Areas – Index of Dissimilarity, 2011

Birthplace	ID
Bosnia and Herzegovina	66.8
Canada	28.3
China	43.6
Croatia	48.9
Egypt	41.7
Fiji	35.3
Former Yugoslav Republic of Macedonia (FYROM)	79.2
Germany	22.0
Greece	48.0
Hong Kong (SAR of China)	38.2
India	37.6
Indonesia	36.0
Iraq	65.8
Ireland	23.4
Italy	43.2
Japan	57.6
Korea, Republic of (South)	55.3
Lebanon	57.4
Malaysia	32.9
Netherlands	21.8
New Zealand	37.9
Papua New Guinea	44.1
Philippines	29.7
Poland	36.3
Singapore	38.9
South Africa	37.1
Sri Lanka	38.6
Thailand	34.0
Turkey	66.2
United Kingdom, Channel Islands and Isle of Man	24.4
United States of America	25.9
Vietnam	44.7

Source: ABS 2011 Census

some longer standing Southern European groups still have relatively high I_Ds. However even MES origin groups, such as those from New Zealand and the UK, have moderate I_Ds suggesting some degree of spatial concentration, which reflects their concentration in particular non-metropolitan niches such as mining communities. Non-metropolitan I_Ds are somewhat higher than metropolitan I_Ds for almost all overseas-born groups. Overseas-born, especially CALD groups, are concentrated because of the importance of networks in non-metropolitan area settlement, concentrations of particular job opportunities, chain migration and, for some groups, deliberate settlement policy.

Temporary Migrants in Non-Metropolitan Areas

A profound change in Australia's immigration system since the mid-1990s has been an increase in temporary migration. On 30 June 2012 there were 1,046,839 persons temporarily present in Australia (DIAC, 2012a) and, until the onset of the Global Financial Crisis, these numbers were increasing by 15 percent per year. Since DIAC (2009) reports that 64 percent of groups stay in Australia longer than 3 months, where they go when they arrive in Australia has an impact on population distribution, housing and labour markets, and demand for services. One major category are Long Stay Temporary Business Entrants (Visa Category 457) who numbered a record 110,280 in 2007-08. Although the numbers declined a little in 2008-09 (101,280), they recovered to a new record number of 140,769 in 2011. These migrants are restricted to the top three skill categories, are nominated by an employer and may stay in Australia up to four years. They are more concentrated in major cities than permanent migrants; some 51 percent of all 457s coming in 2001-03 went to Sydney and 84 percent went to Australia's five largest cities (Khoo *et al.* 2003). In 2002 a regional version of the 457 visa was introduced with a number of 'concessional arrangements ... to reflect the skill needs of regional Australia' (DIMA 2007: 46). These concessions included a lower minimum level of skill and salary than was the case for the standard 457 program, and migrants needed to be endorsed by relevant regional certifying bodies, be at local wage levels and in places where no employable locals were available. The numbers of regional 457s grew quite rapidly but became the subject of controversy because of accusations that some employers used the visa to undercut the wages and conditions of Australian workers in regional areas, especially in the abattoir industry, resulting in tightened regulations. Although 457s are disproportionately concentrated in major cities they are increasingly important in filling job vacancies in regional areas, especially regional cities and mining regions. Temporary skilled migrants are of great significance in regional areas and include doctors and other health personnel.

By 2009-10 temporary skilled migrants were still strongly concentrated in capital cities but with some important concentrations in non-metropolitan areas, especially the mining areas in Western Australia and Queensland. That provoked debate in 2012 about the proposals of some mine operators to bring in significant numbers of temporary skilled migrants. Other areas of concentration of 457s include the regional steel manufacturing city of Whyalla in South Australia. The steel company was having great difficulty in attracting engineers to the city and developed a relationship with a recruiting agency for engineers from South Africa. The 457 visa is thus increasingly being used to fill skill shortages in regional areas.

The largest category of temporary residents is overseas students who numbered 317,897 in 2008. Again they are strongly concentrated in major mainland cities with the most universities and other tertiary education institutions. Regional centres with universities, like Ballarat and Bathurst, are making a substantial effort to attract students both to contribute to the local economy as students and in the hope that they will subsequently become local permanent residents.

One category of temporary migration which has increased in scale over the last decade and involved regional Australia is Working Holiday Makers (WHM). This program involves 'the temporary entry and stay of young people wanting to combine a holiday in Australia with the opportunity to supplement travel funds

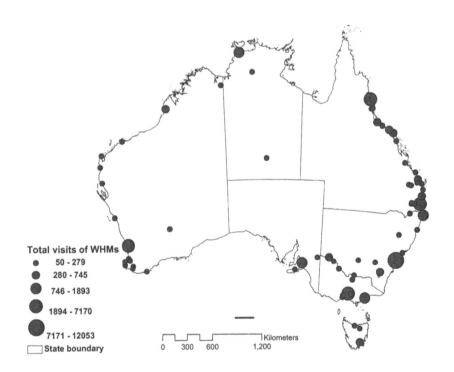

Figure 4.5   Major Localities which WHMs Visited, 2008

Source: Tan *et al*. 2009: 70

through incidental employment' (DIMA 2007: 64). WHM can stay for a year and work in a single job for up to 3 months. They are especially involved in the hospitality and agricultural industries, where many jobs are outside major cities, and have become critically important in providing seasonal harvest labour in horticultural, irrigated fruit growing and grape harvesting activities (Hugo 2001). Indeed they have been so significant that since late 2005 WHM who had undertaken seasonal work in regional Australia for three months were eligible to apply for a second 12 month visa. In 2010-11 some 130,612 WHM visas were granted, an increase of 21 percent on the previous year and a doubling since 2003-04. They have become an important seasonal element in the population of particular regional communities (Figure 4.5) which are much less metropolitan than is the case for other immigrant groups, despite a strong concentration in coastal areas. Even so, an earlier survey of WHMs reported 42 percent spending some time working in Sydney alone (Harding and Webster 2002), hence large cities are also significant for WHMs.

In 2008 the Australian government relented to pressure from the agricultural industry and introduced the Pacific Seasonal Worker Pilot Scheme (PSWPS) aimed at allowing Pacific Islanders to fill seasonal labour shortages in the horticultural industry and modelled on an existing scheme in New Zealand. Despite the scheme becoming permanent in 2011, the take-up has been fairly low, with 1,100 workers having been deployed by March 2012. It is apparent that growers rely heavily on the WHM but the PWSPS may become more important as the scheme becomes better known.

Refugee Settlement in Regional Australia

Settlement of refugees in non-metropolitan Australia has a long history. In the immediate postwar years Australia initiated its first substantial organised immigration of non-British settlers – Displaced Persons (DPs) – from Eastern Europe (Kunz, 1988; Price, 1990). One condition was that they were obliged to work for their first two years in Australia in a location identified by the government. Many such places were in non-metropolitan areas which were suffering significant labour shortages such as the Snowy Mountains Scheme in New South Wales and Victoria, and the Hydro Electric Commission in Tasmania, but also in isolated rural communities and even railway sidings needing unskilled workers (Hugo 1999; Kunz 1988). They also included expanding industrial provincial cities such as Wollongong, Geelong and Newcastle. Many DPs were highly qualified but unable to use their particular skills or have their qualifications recognised (Kunz 1975; 1988). Many gravitated to the major cities after their two years were up but significant numbers remained in provincial communities like Cooma in New South Wales, the centre of the Snowy Mountains Scheme, and elsewhere (Figure 4.2). While DP migration subsided in the mid 1950s subsequent attempts were made by governments to settle refugee-humanitarian immigrants in non-metropolitan areas. With the influx of Vietnamese refugees in the 1970s and 1980s, the federal government worked with NGOs for settlement in regional centres like Whyalla in South Australia (Viviani *et al.* 1993). Backing these initiatives local NGOs had indicated that they would support and assist refugee settlement in those areas and the government provided some additional support. However, many of the refugees also gravitated to metropolises like Sydney and Melbourne, with significant Vietnamese communities and the opportunity to access social and economic support. Regional centres usually lacked both formal multicultural services and substantial ethnic communities to provide informal support.

The most significant government efforts to facilitate refugee-humanitarian newcomers settling in non-metropolitan areas, however, have come in this century. In 2003 a Department of Immigration and Citizenship (DIAC) Review of Settlement Services for Migrants and Humanitarian Entrants

Figure 4.6   Australia: Settlement of Refugee-Humanitarian Settlers Outside Capital Cities, 1996-2011

Source: DIAC

recommended that more refugees be settled in non-metropolitan areas, and DIAC developed a new approach for identifying and establishing regional locations for humanitarian settlement in 2005 (DIAC 2009). This approach focused on so-called 'unlinked migrants' or refugee-humanitarian settlers who did not have established local family ties. A quarter of refugee-humanitarian settlers had no family members living in Australia, hence were most reliant on formal support services and the wider community. It was considered that, with support from government and local communities, they would be able to as effectively adjust to life in Australia in regional areas as they could in major cities. The Department set up a number of criteria to identify particular regional areas which would be selected for directed settlement of humanitarian immigrants: a population of more than 20,000, existing migrant communities, evidence of community acceptance of immigrants, an accessible location, and the availability of appropriate employment opportunities and service infrastructure (especially health and education). Communities did not have to meet all these criteria but did need to meet most in order to qualify, and communities selected by DIAC received some resources to support services for the settlers. The program has had a significant impact on patterns of refugee-humanitarian settlement in Australia, with the proportion of refugee-humanitarian settlers initially moving to communities outside the capitals quadrupling to one in five over the last decade (Figure 4.6).

An Increasing Government Role in the Geography of Migrant Settlement

The 2001-06 intercensal period saw for the first time in several decades a reversal of the trend of increasing concentration of new migrant arrivals in capital cities, with government policy playing a role in this change. During the early post-war period, Australian immigration policy was overwhelmingly concerned with shaping the scale and composition of the immigration intake with few attempts to influence where immigrants settled after arrival. It was not until the mid-1990s that the government considered major initiatives to substantially shape where immigrants settle. The sustainability of rural and regional communities became an important national issue with the establishment of a federal government department on regional development and the initiation of various programs to facilitate regional development. Simultaneously, states which were lagging economically, like South Australia, were pressing for immigration to assist their economic development.

In 1996 the annual meeting of Commonwealth, State and Territory Ministers for Immigration and Multicultural Affairs established a working party to examine ways in which a higher proportion of migrants might settle in regional Australia and in states which were lagging economically, and a number of initiatives followed. A State Specific and Regional Migration Scheme (SSRM) was initiated in 1996 to attract immigrants to areas which were receiving small intakes. Several visa categories were added to the scheme and a range of modifications made over subsequent years to enable employers, state and local governments and families in designated lagging economic regions to sponsor immigrants without them having to fully meet the stringent requirements of the Australian Points Assessment Scheme. That scheme focused on skills, restricting most SSRM visa categories to people who narrowly missed reaching the high pass threshold of the Points Assessment Scheme. The new regional program (The Regional Sponsored Migration Program, RSMS) sought to offer:

> greater flexibility for employers by recognising that labour market conditions are complex and that labour supply is limited in many regional and remote communities. Some concessions offered through the RSMS include the capacity for employers to nominate a greater range of occupations and a lower salary requirement compared to other temporary and permanent employer sponsored visas (DIAC, 2012b, 17).

Some categories required settlers to live in a designated area as a temporary resident for three years, after which their degree of adjustment would be assessed and they would be given permanent residence, and the right to settle anywhere in Australia. Foreign students who studied in an institution in a designated area got five bonus points in the Points Assessment Test. A 'Regional 457' (Long Term Business Migrants) visa was also developed, giving regional bodies a greater role in supporting sponsorships in regional Australia. It allowed them to grant exceptions from the gazetted minimum skill and salary requirements for positions nominated under temporary business visas in regional and low population growth areas.

The growing significance of RSMS was evident with numbers trebling between 2006-07 and 2010-11 as employers increasingly recognised this means of acquiring skilled workers (Table 4.6). Queensland (28 percent), South Australia (18 percent), Victoria (14 percent) and Western Australia (14 percent) are the main users of the program (DIAC, 2012b, 18). The occupations of those selected (Table 4.7) reflect skill shortages in regional areas, especially in medical professions, tourism, skilled artisans, meat processing and farming. Regional authorities and employers became increasingly aware of the RSMS. In South Australia, for example, the state government appointed Migration Officers to each of the Regional Development Boards to facilitate recruitment and settlement of settlers in regional parts of the state. This was because at no time since Federation a century earlier have regional organisations and employers been so involved in immigration policy and operations.

One strategy that DIAC developed to meet labour shortages in non-metropolitan areas was Labour Agreements enabling a business to employ specialised overseas workers when no other visa program meets the employer's needs. Labour Agreements are most commonly used by businesses seeking semi-skilled labour or by those in the on-hire and meat industries. They are a form of negotiated

Table 4.6   Australia: Main RSMS Occupations – Primary Applicants, 2009-10[1]

Occupation	2009-10	2010-11
Registered nurse	717	562
Cook	228	655
Motor mechanic	166	217
Welder (first class)	149	266
Chef	111	208
Metal fabricator	108	125
Slaughter person	106	135
General medical practitioner	74	117
Farm overseer	73	125
University lecturer	60	79
Agricultural technical officer	57	108
Restaurant and catering manager	52	85
Other	1,866	2,072
Total	**3,979**	**4,764**

[1] Occupation data is only available for primary applicants within the Skill Stream

Source: DIAC, 2012b, 18

contract to employ overseas workers when workers cannot be found in the local labour market. Labour Agreements are most usually applied in regional areas and in 2011 there were 123 with 84 under negotiation (DIAC 2012b: 19). They involved substantial negotiation and were strongly criticised by unions, especially in the light of increasing unemployment among manufacturing workers due to closure of enterprises in southeastern Australia in 2011-12. Labour Agreements were identified by DIAC as a way to meet the substantial labour demands created by the expansion of the mining industry, especially in Western Australia where it was difficult to attract workers from the east coast.

Drivers of Immigrant Settlement in Regional Australia

The involvement of government policy is part of a complex set of factors which lay behind increasing settlement of skilled immigrants and refugee-humanitarian entrants in regional Australia. Availability of jobs is essential for immigrant settlement, and labour shortages and demand in particular sectors in particular regions have been important drivers. A number of dimensions to this demand include shifts in global demand for regional based industries – agriculture, mining and tourism, restructuring of specific industries in Australia, the heavy outmigration of young Australians from regional areas and the effects of cumulative causation and chain migration.

The Australian economy has experienced significant structural change over recent decades with declining employment in agriculture and manufacturing and increases in mining and services. Employment in agriculture, forestry and fishing in Australia declined from 380,900 in 1981 to 249,827 in 2011 (see Chapter 10). However, the proportion of overseas-born in this sector increased from 10 to 14 percent over the period. While migrants are still underrepresented compared with the Australia-born, their involvement is increasing, although it is still mainly in the intensive horticulture and irrigated agriculture sectors in the Murray-Darling Basin and in the immediate hinterlands of large cities. This trend is especially significant in light of the Head of the Australian Treasury arguing that the percentage of the Australian workforce engaged in agriculture will increase from the current 2 percent of GDP to 5 percent in 2050 due to increasing global and regional food security issues and demand for quality food from Asia's burgeoning middle classes (Parkinson 2012).

In the United States a key area of increasing employment of immigrants outside the major gateway areas has been in restructured food processing industries. As Leach and Bean (2008: 55) point out: 'the industry reduced production costs by relocating to rural areas and deskilling production processes while simultaneously working to weaken labour unions, thus increasing the need to recruit immigrant labour to reduce labour costs'. Some 60 percent of food processing in the United States is now in rural areas (Kandel 2009). In Australia this pattern is most evident in the meat processing and abattoir industries which have been decentralised to regional communities creating a significant new demand for workers. The jobs in the industry are low paid, low status, manual jobs eschewed by local Australians creating a demand for immigrant labour. In the United States: 'The presence and expansion of poorly paid jobs that are difficult, dirty and sometimes dangerous, in small towns and rural areas is a common thread in many "new destination" areas' (Hirschman and Massey 2008: 8). The meat processing industry in non-metropolitan Australia has been a major employer of immigrants. Refugee-humanitarian settlers have been an important source of such labour while some meat processing operations have brought in workers from China under labour agreements.

While the numbers of Australians employed in the mining industry is still small (176,560 in 2011, less than 2 percent of the workforce) it is the most rapidly growing industry sector, expanding by 65 percent over the 2006-11 period. The immigrant engagement in the industry is around the average for the entire workforce (23 percent) but it has increased with 40,894 overseas-born employed in 2011. Mining industry operations are almost exclusively located in regional areas but only 63 percent of those working in mining in 2011 lived outside capital cities. This is a function of both corporate and administrative activity in mining being located in cities and the 'fly-in-fly-out' and 'drive in-drive out' phenomenon (ABS 2008; see Chapter 7). Nevertheless, the overseas-born living outside the capitals working in mining increased by 0.4 to 16 percent over the 2006-11 period. In discussions of labour shortages in regional mining activity, especially in Western Australia, immigration has loomed large (ABS 2008). There has been strong opposition from unions to the proposal of some mining companies to use Labour Agreements to bring in overseas workers. A third sector of the economy that has a strong regional presence and, until recently, was expanding rapidly, is tourism. Recently the industry has been hit by the high Australian dollar but it has been a major beneficiary of the Asian economic boom of the last two decades with seven of the ten most valuable Australian tourist markets being in Asia (TRA 2012). Immigrants have long been an important element in the regional tourist workforce with language factors being of some importance, especially for areas attracting large numbers of Asian tourists (Bell and Carr 1994). Other areas of labour demand which have been important for immigrants have been the expanding regional universities and health and education services. Health services are of increasing significance with ageing of the regional populations being exacerbated by significant levels of retirement migration, especially to coastal areas.

One key element in the explanation of increased immigrant settlement in non-metropolitan areas, however, is not so much related to expansion of new employment opportunities but to the offsetting of longstanding outmigration of local young people from non-metropolitan areas (Hugo 1974; see Chapter 2). Critically this offsetting has brought in younger working and family age groups, hence overseas-born net migration gains have a major impact on regional communities through, firstly, meeting important labour shortages, both high and low skilled, which are crucial for local economic sustainability. Secondly, since they are often young families they play an important role in the social sustainability of those communities by creating demand for local goods and services, particularly in health and education. In addition they play a crucial role in the social life of communities in volunteering, participation in sport and other organisations.

Between 2006-2011 patterns of net migration in regional Australia demonstrate a net loss in the young adult ages for the total population, but the pattern for the overseas-born (Figure 4.7) shows a clear offsetting impact. There are net gains in all but the very oldest age groups but they are most marked in the 20s and 30s groups. While there was a net loss of 106,835 Australia-born aged 15-24 between 2006 and 2011 there was a net gain of 28,994 overseas-born in those age groups. Hence the inmigration of overseas-born negated over a quarter of the net loss of young Australia-born in the 15-24 age group. In the 30-45 prime working age

(a) Total

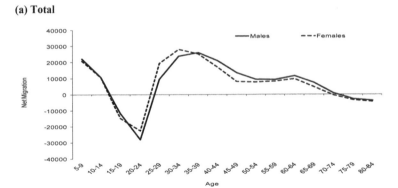

(b) Australia-born

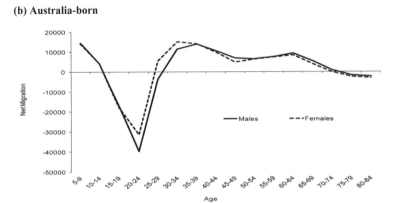

(c) Overseas-born

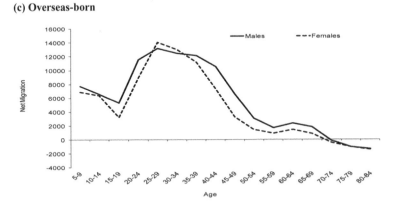

Figure 4.7   Australia: Rest of State Migration Profile, 2006-2011

Source: ABS 2006 and 2011 Censuses

groups the net gain of the overseas-born (66,636) is almost as large as the net gain of the Australia-born (74,345). The invaluable role of overseas migration in being a 'replacement factor' and in providing a regional workforce is slowly becoming evident.

Social networks and institutions of mutual support lead to immigrant concentrations developing through pioneer migrants supporting and encouraging subsequent migration of family and friends. As Hirschman and Massey (2008: 10) point out: 'Each pioneer immigrant commonly creates the potential for additional immigration through network-driven processes of cumulative causation, and eventually for the creation of satellite settlements in nearby towns where immigrant niches can be reproduced'. The latter has clearly been the case, for example, in the Southeast region of South Australia where small immigrant concentrations have developed in a number of localities as a result of a major concentration in the regional centre of Naracoorte. Chain migration has a long history in the development of concentrations of immigrants in non-metropolitan Australian communities (Hugo 1975). For more than half of recent immigrant settlers in regional communities, family and friends were the main source of information about the community in which they settled (Collins 2009: 29). Hence as the numbers of immigrants settling in regional areas increases, the potential for chain migration also increases.

Immigrants' Experience in Regional Australia

An emerging body of research relates to the adjustment of new immigrants in non-metropolitan communities and their impact on those communities (Wulff *et al.* 2008; Collins 2009; Wulff and Dharmalingam 2008; Forrest and Dunn 2013; Taylor-Neumann and Balasingam 2013; Hugo 2008; Jordan *et al.* 2011). Immigrants are evidently filling labour shortages in particular niche markets. In some areas of intensive agriculture, harvest labour is strongly dependent on Working Holiday Makers (Tan *et al.* 2009). In remote mining areas the resident populations, as opposed to fly-in-fly-out workers, have a strong representation of recent migrants. The proportions of workers in Agriculture, Forestry and Fishing that were overseas-born increased from 12 to 14 while for Mining it increased from 21 to 23 percent between 2006 and 2011.

Recent immigrants have been especially significant in regional food processing. This is similar to the experience in the United States where, as Sanderson and Painter (2011: 403) point out:

> Over the past 30 years, the U.S. food-processing industry has been restructured from a predominately urban enterprise with relatively high rates of unionization and competitive wages into a rural-based industry with very high rates of firm consolidation, sales concentration, precarious forms of employment, higher turnover rates, declining wages, and an expanding Hispanic immigrant workforce.

Not all of these trends are apparent in Australia but some are, especially in meat processing. More than three quarters of the fifty largest abattoirs in Australia, which have over 20,000 workers, are now located outside capital cities, as those in larger cities like Sydney have gradually closed down. The 'dirty, dangerous and difficult' work in abattoirs has discouraged local workers. So too have agricultural vacancies been filled by new migrants, notably in Griffith (NSW), where Indian Sikhs, Turks and Pacific Islanders work in fruit picking, farming and agricultural processing (Jordan *et al.* 2011). Critical shortages of workers, that have discouraged business expansion, have partly been met through engaging immigrant workers.

Refugee-humanitarian settlers have been an important element in meeting the shortage of meat processing workers in regional Australia, a parallel to the Hispanic chain migration in the United States. Refugee-humanitarian settlers have become important to the sustainability of several abattoirs in rural Australia. Thus Afghan refugee-settlers have been valuable in the New South Wales town of Young where many work in meat processing, with the Federal government policy of settling refugees in non-metropolitan areas assisting in their initial recruitment. The workers and their families have made a substantial economic contribution, quite in contrast to prevailing national discourses which depict refugees as social problems and economic burdens (Stilwell 2003: 247). Prejudice against migrants, especially Muslims, nonetheless remains, there and elsewhere (Jordan *et al.* 2011). In some cases owners of regional abattoirs have also brought in 457s to meet labour shortages. This has been the case in Port Wakefield, Murray Bridge and Naracoorte in South Australia where workers are sourced from China. All these towns have had limited previous settlement from CALD countries.

However, it is not only in meeting shortages of 'low pay, low skill' workers that international migration is having an impact. New settlers are adding significantly to the skilled human capital in local communities (Collins 2009). Moreover the migrant population in regional Australia, when compared with the Australia-born, have significantly higher levels of education, especially among the most recent arrivals (Massey and Parr 2012). While in the past overseas-born groups have experienced relative socio-economic disadvantage compared with the whole of Australia, the gap has closed as a result of the SSRM schemes. Consequently increasing migration to regional and rural areas has a number of benefits, including filling skill shortages, and helping to reinvigorate regional economies through the influx of highly educated and skilled groups. Longer-term economic benefits, however, are dependent on migrants being retained in regional communities. Under several of the SSRM schemes settlers are required to remain in the communities of initial settlement for their first two years in Australia but then are free to go elsewhere in Australia. Social connectedness is crucial to ensuring that immigrants remain in their areas of settlement. Among the factors enhancing social connectedness include coming from a Mainly English-Speaking origin, having children and having assistance from sponsors upon arrival (Wulff and Dharmalingam 2008). Service provision is as important, with Sudanese migrants in the South Australian town of Murray Bridge being satisfied with housing and employment but frustrated by lack of access to tertiary training and thus seeking to move to Adelaide (Taylor-Neumann and Balasingam 2013).

Traditional stereotypes of non-metropolitan areas depict them as less progressive and more conservative than their city counterparts. As Forrest and Dunn (2013: 1) point out: 'Rural areas have also been perceived as "white" landscapes where cultural diversity and even ethnicity is rarely "seen"'. In South Australia, in the absence of ethnic diversity, there were lower levels of tolerance and more conservative attitudes outside Adelaide than within it. However, there were significant variations in racist attitudes within non-metropolitan areas depending on the particular mix of socio-demographic and population diversity. Elsewhere in South Australia, in some communities Afghani, Chinese and Sudanese families have been welcomed while in others overt racist attitudes have been evident. There is, as Forrest and Dunn (2013: 8) point out, a need for deeper understanding of rural attitudes toward new immigrants if the settlement of these groups is to be facilitated.

Conclusion

Immigrant settlement in regional Australia has a long history but recent migration differs from earlier flows in a number of ways. Firstly it has been diverse in terms of birthplace groups and involving a wider range of visa categories – skilled, family, Working Holiday Makers, refugees and students. Secondly it has been more spatially dispersed than earlier decades with more settlement in the wheat-sheep belt. It also represents a small but nevertheless significant reversal of the consistent postwar trend of increasing concentration of immigrant settlement in the gateway cities. What of the future? Jordan *et al.* (2011: 260) argue that while the 'regionalisation' of Australian immigration policy has only achieved a small reversal of the post-war trend in increasing concentration of immigrants in Australian capital cities: 'the numbers are critical to regional and rural Australia and represent a turning point in Australian immigration history'. The role of international migration in non-metropolitan Australia is likely to increase in importance over the next decade.

Although immigrants to Australia will continue to concentrate in large 'gateway cities', they are playing an increasing role in regional Australia. Their impact is nevertheless considerable and, as in the United States, there is a new diversity in immigrant settlement (Hirschman and Massey 2008: 3). The extent to which the new settlers will remain and be encouraged to stay in those communities in the longer term, and their impact and role in these communities, will depend on the availability of formal and informal support services, employment, education and housing opportunities, discrimination and the role of chain migration. What the new migration has achieved is to offset rural outmigration and link rural and regional Australia to a much wider world.

References

Australian Bureau of Statistics (ABS), 2006, *Census Dictionary: Australia 2006*, Canberra: ABS.

Australian Bureau of Statistics (ABS), 2008, *Australian Social Trends 2008*, Canberra: ABS.

Australian Bureau of Statistics (ABS), 2012, *Year Book Australia, 2012 – Migrant Farmers*, Canberra: ABS.

Bell, M. and Carr, R., 1994, *Japanese Temporary Residents in the Cairns Tourism Industry*, Canberra: AGPS.

Borrie, W.D., 1954, *Italians and Germans in Australia. A Study of Assimilation*, Melbourne: Cheshire.

Collins, J., 2009, *Attraction and Retention of New Immigrants in Regional and Rural Australia; Report on PEP Leave Spring Semester 2008*, Sydney: University of Technology Sydney.

Department of Immigration and Citizenship (DIAC), 2009, *Refugee and Humanitarian Issues: Australia's Response*, Canberra: AGPS.

Department of Immigration and Citizenship (DIAC), 2012a, *Immigration Update – July to December 2011*, Canberra: AGPS.

Department of Immigration and Citizenship (DIAC), 2012b, *Population Flows: Immigration Aspects 2010-11 Edition*, Canberra: AGPS.

Department of Immigration and Multicultural Affairs (DIMA), 2007, *Population Flows: Immigration Aspects*, 46, Canberra: DIMA.

Forrest, J. and Dunn, K., 2013, 'Cultural Diversity, Racialisation and the Experience of Racism in Rural Australia: The South Australian Case', *Journal of Rural Studies*, 30: 1-9.

Green, A., de Hoyos, M., Jones, P. and Owen, D., 2009, 'Rural Development and Labour Supply Challenges in the UK: The Role of Non-UK Migrants, *Regional Studies*, 31: 641-657.

Harding, G. and Webster, E., 2002, *The Working Holiday Maker Scheme and the Australian Labour Market*, Melbourne Institute of Applied Economic and Social Research, University of Melbourne.

Hirschman, C. and Massey, D.S., 2008, 'Places and Peoples: The New American Mosaic', in D.S. Massey (ed.), *New Faces in New Places: The Changing Geography of American Immigration*, New York: Russell Sage Foundation, 1-22.

Hugo, G.J., 1974, 'Internal Migration and Urbanization in South Australia', in Burnley, I.H. (ed.), *Urbanization in Australia – The Post War Experience*, Cambridge: Cambridge University Press, 81-98.

Hugo, G.J., 1975, 'Postwar Settlement of Southern Europeans in Australian Rural Areas: The Case of Renmark', *Australian Geographical Studies*, 13: 169-181.

Hugo, G.J., 1994, 'The Turnaround in Australia: Some First Observations from the 1991 Census', *Australian Geographer*, 25(1): 1-17.

Hugo, G.J., 1999, 'Regional Development Through Immigration? The Reality Behind the Rhetoric', *Department of the Parliamentary Library Information and Research Services Research Paper No. 9*, 1999-2000, Canberra: AGPS.

Hugo, G.J., 2002, 'From Compassion to Compliance?: Trends in Refugee and Humanitarian Migration in Australia', *GeoJournal*, 56: 27-37.

Hugo, G.J., 2004, 'Australia's Most Recent Immigrants 2001', *Australian Census Analytic Program*, Canberra: Australian Bureau of Statistics.

Hugo, G.J., 2008, 'Australia's State Specific and Regional Migration Scheme: An Assessment of Its Impacts in South Australia', *Journal of International Migration and Integration*, 9: 125-145.

Hugo, G.J., 2011, 'Changing Spatial Patterns of Immigrant Settlement', in M. Clyne and J. Jupp (eds), *Multiculturalism and Integration – A Harmonious Combination*, Canberra: ANU E Press, 1-40.

Hugo, G.J. and Bell, M., 2000, *Internal Migration in Australia 1991-1996 Overview and the Overseas-Born*, Canberra: AGPS.

Jordan, K., Krivokapic-Skoko, B. and Collins, J., 2011, 'Immigration and Multicultural Place-making in Rural and Regional Australia', in G. Luck, D. Race and R. Black (eds), *Demographic Change in Australia's Rural Landscapes: Implications for Society and the Environment*, Dordrecht: Springer, 259-280.

Jupp, J., 1993, 'Ethnic Concentrations: A Reply to Bob Birrell', *People and Place*, 4(4): 51-52.

Kandel, W., 2009, 'Recent Trends in Rural-based Meat Processing', *Presentation at "Immigration Reform: Implications for Farmers, Farm Workers and Communities" Conference*, Washington DC, May 21-22.

Khoo, S., Voigt-Graf, C., Hugo, G. and McDonald, P., 2003, 'Temporary Skilled Migration to Australia: The 457 Visa Sub-Class', *People and Place*, 11(4): 27-40.

Kunz, E.F., 1975, *The Intruders: Refugee Doctors in Australia*, Canberra: Australian National University Press.

Kunz, E.F., 1988, *Displaced Persons: Calwell's New Australians*, Canberra: Australian National University Press.

Leach, M.A. and Bean, F.D., 2008, 'The Structure and Dynamics of Mexican Migration to New Destinations in the United States', in D. Massey (ed.), *New Faces in New Places: The Changing Geography of American Immigration*, New York: Russell Sage Foundation, 51-74.

Massey, D.S. (ed.), 2008, *New Faces in New Places: The Changing Geography of American Immigration*, New York: Russell Sage Foundation.

Massey, S.J.L. and Parr, N., 2012, 'The Socio-Economic Status of Migration Populations in Regional and Rural Australia and Its Implications for Future Population Policy', *Journal of Population Research*, 29: 1-22.

Oliva, J., 2010, 'Rural Melting-Pots, Mobilities and Fragilities: Reflections on the Spanish Case', *Sociologia Ruralis*, 50: 278-295.

Parkinson, M., 2012, 'Challenges and Opportunities for the Australian Economy', Speech to the John Curtin Institute of Public Policy, Breakfast Forum, 5 October. http://www.treasury.gov.au/PublicationsAndMedia/Speeches/2012/Challenges-and-opportunities-for-the-Aust-economy.

Price, C.A., 1963, *Southern Europeans in Australia*, Melbourne: Oxford University Press.

Price, C.A., 1990, 'Australia and Refugees, 1921-1976', in *National Population Council*, Refugee Review – Volume 2 – Commissioned Reports, Canberra: National Population Council.

Sanderson, M. and Painter, M., 2011, 'Occupational Channels for Mexican Migration: New Destination Formation in a Binational Context', *Rural Sociology*, 76: 461-480.

Spoonley, P. and Bedford, R., 2012, *Welcome to our World? Immigration and The Reshaping of New Zealand*, Auckland: Dunmore Publishing.

Stilwell, F., 2003, 'Refugees in a Region: Afghans in Young, NSW', *Urban Policy and Research*, 21(3): 235-248.

Tan, Y., Richardson, S., Lester, L., Bai, T. and Sun, L., 2009, 'Evaluation of Australia's Working Holiday Maker (WHM) Program', Adelaide: Flinders University National Institute of Labour Studies.

Taylor-Neumann, N. and Balasingam, M., 2013, 'Migratory Patterns and Settlement Experiences of African Australians of Refugee Background in Murray Bridge, South Australia', *Australian Geographer*, 44: 161-176.

TRA (Tourism Research Australia), 2012, 'Tourism Industry Facts and Figures at a Glance', Canberra: Department of Resources, Energy and Tourism.

Viviani, N., Coughlan, J. and Rowland, T., 1993, *Indo-Chinese in Australia: The Issues of Unemployment and Residential Concentration*, Canberra: AGPS.

Wulff, M. and Dharmalingam, A., 2008, 'Retaining Skilled Migrants in Regional Australia: The Role of Social Connectedness', *Journal of International Migration and Integration*, 9: 147-160.

Wulff, M., Carter, T., Vineberg, R. and Ward, S., 2008, 'Attracting New Arrivals to Smaller Cities and Rural Communities; Findings from Australia, Canada and New Zealand', *Journal of International Migration and Integration*, 9: 119-124.

Chapter 5

'They have no concept of what a farm is': Exploring Rural Change through Tree Change Migration

Danielle Drozdzewski

Introduction

While much discussion of population change in rural and regional Australia centres on population loss, and especially the flight of the young, there has also been a distinct reverse trend: counterurbanisation or the population turnaround (Burnley 1988, Burnley and Murphy 2002, Walmsley *et al.*1996). Indeed to a large extent Australian cities had captured their largest proportion of the national population by the 1980s. Thereafter people slowly began to move into 'perimetropolitan zones' (Hugo 1994) on the fringes of the largest cities, and somewhat further away in a process of sea change that transformed many small coastal towns such as Noosa and Byron Bay (Burnley and Murphy 2004) and was seemingly sanctified in the popular ABC television series *Sea Change* that ran through the late 1990s. Other parallel patterns of mobility included 'welfare migration', associated with lower income households moving out of the city in search of cheaper housing, and their converse, 'urban refugees' who had no wish to give up urban amenities and employment, but sought a more rural life within commuting distance of the city, and who have taken up 'acreage' west of Brisbane, in the Yarra ranges east of Melbourne and around Windsor and Richmond northwest of Sydney. The population turnaround was largely driven and facilitated by a combination of greater affluence and longevity, mobility, housing affordability and environmental amenity in regional areas – and some prospect of a new lifestyle – a broadly-based amenity led migration which for some meant 'downshifting' (Connell and McManus 2011, Argent *et al.* 2011). There was one other important population shift – a tree change propelled by many of the same factors that influenced sea change – but took people inland and, in some cases, to places where councils were actively seeking workers and families. This chapter examines this particular process and the changes that it has wrought for inland Australia.

That councils in many parts of Australia have actively sought population growth and considered a range of inducements and strategies to stimulate migration certainly suggests that there are disadvantages to regional residence.

Ragusa (2011a: 75) has argued that part of metropolitan Australians' 'reluctance to move to the country has been a hesitation to give up urban lifestyle amenities'. In Australia – where over two thirds of the population live in a capital city – a move out of a major capital city has been equated with 'locational disadvantage' (Ware *et al.* 2010). Disparities in the provision of services and amenities outside capital cities have intensified following the application of neoliberal policies in the Australian economy in the 1990s. In rural areas this has meant an almost certain reduction in access to health and specialist services, a substantial reduction in educational and employment opportunities, poor access to public transport, and a reduced range of entertainment options. Counteracting such disadvantages poses problems.

In Europe, prospective migrants can undertake rural relocation while still being within reasonable commuting distance of major service centres. Australians undertaking tree change usually travel further than their northern hemisphere counterparts; defining features of contemporary internal migration in Australia reflect the vast distances between regional towns and state capitals, and the growing unfamiliarity of metropolitan populations with regional Australia. Accounting for and understanding the dual role of distance and access to services aids an appreciation of why the more dominant migration pathway has been rural to urban, though accompanied by glib media rhetoric about Australia's dying rural and inland towns (Collits 2000). Winchester and Rofe (2005: 269) contend that this discursive positioning places the regional and rural as a 'place apart' from metropolitan Australia, one frequently connected with, if no longer dependent on, agriculture (Beer *et al.* 2003, Gray and Lawrence 2001). In this sense, places outside capital cities are commonly associated with visual aspects of the rural idyll – rolling countryside, pasture, homesteads, farming (see Chapter 11). These rural imaginings are important to people's decisions to move and, as this chapter will show, to how migrants and local people conceptualise regional and rural change.

International literature suggests that central to tree change migration is the search for the 'rural idyll' and the rejection of the 'vices' and anomie of city life (Benson and O'Reilly 2009). Rural life is problematically positioned as 'uncomplicated, innocent, [and a] more genuine society in which 'traditional values' persist and lives are more real' (Little and Austin 1996: 102). Idyllic images of the rural are coupled with metaphors of a more wholesome lifestyle. Tree change migration is thus primarily positioned as 'lifestyle' based and the converse of conventional economic drivers of migration (Bohnet and Moore 2011, Halfacree and Rivera 2012). Salt (2009) has contended that tree change migration is the move you make when you either do not like or cannot afford to move to the beach. However economic and lifestyle factors are inextricably (and commonly) linked, so that prospective tree changers do not necessarily dislike urban areas, but are spurred to move by rising housing costs and the greater affordability of regional housing (Butt 2011, Connell and McManus 2011). Tree change migrants are as likely to be discouraged by urban life as encouraged by rural and small town life.

This chapter examines how recent tree change migration has influenced landscapes and livelihoods. Research in sea change locations has questioned whether new migrants actually stimulate economic growth and lead to an increase in the provision of services and facilities. Migrants may not necessarily engender 'improved socio-economic outcomes for local populations' and can exacerbate environmental problems (Gurran *et al.* 2005, Gurran 2008) and inflate property prices (Bohnet and Moore 2011) where the 'commodification of rurality' has brought what amounts to 'creative destruction' (Tonts and Greive 2002). At the same time Burnley and Murphy (2004: 129) found a 'demographic-economic multiplier effect', where population growth stimulates service provision. In many tree change locations, providing services to dispersed rural populations has proved difficult, and communities have sought an increased population in order to retain, let alone increase, the range of accessible services, as such services have otherwise tended to decline (Argent 2008, Smailes 2002). Much like sea change migration, tree change migration, in Castlemaine, Victoria, has affected local housing markets, rental stock and housing affordability, and changed both the dynamics of the local community and the town's fabric through the development of new amenities with both positive and negative consequences (Costello 2007). New migration has influenced the provision of goods and services (for example, education, recreation, entertainment and health services), land use changes (from new housing developments) and changes to the regional/rural character and even the imaginings of regional life. Such processes constantly evolve, as new phases of migration influence socio-economic change and the perceptions that the locals and migrants have of regional Australia. While this chapter is focused on migration to the relatively large towns of Orange and Bathurst it draws conclusions from a much wider context.

Charting the Push for Regional Relocation

A central narrative of Australian literature on internal migration has been how to enact a long-standing government-led desire to populate Australia's rural interior and regional locations (Bell 1992, Burnley *et al.* 1980, Beer *et al.* 2003). Such strategies have ranged from land grants made by Colonial officers to free settlers and emancipists, to encourage sheep graziers and pastoralists, to legislation to allow squatters to purchase land they had illegally occupied, and Soldiers Settlement Schemes. After World War Two, agricultural expansion was encouraged by government investment into services such as schools, hospitals, rural banks and post offices, and interventionist pricing aimed at lessening inequitable access to jobs and services in rural Australia. While these strategies and schemes initially attracted people away from major urban centres, there has been an equally long and established trend into and back to the city.

In the 1970s, the Whitlam Australian Labor Party Government's Department of Urban and Regional Development (DURD) embarked on a renewed push towards planned decentralisation (Connell and McManus 2011). Part of that strategy was the development of three large inland growth centres: Bathurst-Orange in New South Wales, Albury-Wodonga on the border with Victoria, and Monarto in South Australia. In 1974, the Bathurst-Orange Growth Centre and the Bathurst-Orange Development Corporation (BODC) were established. A primary BODC target was for the population of the Bathurst-Orange Growth Centre to reach 240 000 by 2006. By 1978, the Corporation boasted a budgetary expenditure of A\$32 million on a combination of housing, residential development, industrial premises and land (BODC 1978). Despite these efforts, population growth in Bathurst and Orange fell short of targets and, as Connell and McManus (2011: 24) contended, towards the 1980 and 1990s the nation that supposedly rode on the sheep's back was cementing its trajectory towards 'a coastal, urban and metropolitan future'. The 'big shift' towards the coast – sea change migration – became more apparent and contributed to regional growth being largely concentrated in the 'coastal zone of south-eastern Australia and peri-metropolitan areas' (Burnley and Murphy 2002: 139, Gurran and Blakely 2007: 114). Archetypal sea changers were seen as retirees, working age migrants, rural gentrifers, and also those priced out of urban housing markets (Burnley and Murphy 2004). Popular coastal locations, such as the Gold Coast, Byron Bay and Coffs Harbour, were 'bulging at their boundaries' (Haxton 2005), and many experienced rapidly increasing housing prices.

Institutional strategies and marketing ploys for inland relocation have again multiplied but on a smaller, usually state, scale. In recent years the principle vehicle for the promotion of regional Australia has been Country Week (later the Foundation for Regional Development), a private sector organisation that has promoted regional Australia (at least in NSW and Queensland) through annual three day Expos, where regional councils and government departments publicised the advantages of regional residence: a rare example of collective place-marketing to promote inland Australian towns, to reverse population decline and fill employment vacancies in booming resource towns (Connell and McManus 2011). Australian state governments, notably Victoria and Queensland, have likewise attempted to promote regional growth. The Victorian campaign *Make it Happen in Provincial Victoria* was launched in 2004 by the state government in partnership with the 48 rural and regional councils in Victoria. It had similar aims to Country Week, notably to correct 'misconceptions' about Provincial Victoria, encourage city dwellers to move there and unite rural and regional Victoria under the banner 'Provincial Victoria'. It was rebranded and launched as 'goodmove' in 2012. The Queensland program, *Blueprint for the Bush,* intended to be a ten year program involving the Queensland Government, the Local Government Association of Queensland and Agforce (an agricultural lobbying group), was launched in June 2006. Strategies in Queensland were rather different since, like Western Australia, the state urgently required mine workers in regional areas. At a rather smaller scale a rentafarmhouse scheme, initiated near Bathurst, has had a little success in renting empty rural houses at peppercorn rents to households willing to move there and renovate them.

The most recent relocation strategy has been Evocities, which began in 2012 when seven NSW regional centres (Albury, Armidale, Bathurst, Dubbo, Orange, Tamworth and Wagga Wagga) began a collective campaign to encourage capital city residents to move to an 'Evocity'. The website argued that

> A move to an Evocity is a move to a quality life with less traffic, a greater lifestyle, an embracing and safe community, and a cheaper, bigger house to raise your family. Living in an Evocity means less time commuting, working and stressing, and more time for you and your family to enjoy the Evocities' beautiful yet diverse natural surroundings – from mountainous ranges to stunning lakes, wine regions and dramatic valleys (Evocities 2012).

The strategy was later assisted by the NSW government providing a Regional Relocation (Home Buyers) Grant, starting in 2011, that would give households a grant of $7,000 to assist them in relocating 'from their metropolitan home to a regional home'. Yet, by October 2011, only 49 families had relocated under this scheme; a major barrier to moving was that families needed to have owned their home in Sydney to be eligible for the grant (Nicholls 2011). While Evocities claim that 505 households and a thousand people had migrated in the first two year period, how much of that was influenced by such strategic marketing policies is unknown. Indeed it is impossible to measure, in any of these contexts, the extent to which marketing actually works or whether regional employment and housing markets operate quite independently.

Orange and Bathurst, New South Wales

Orange and Bathurst have both experienced steady population growth in the last five years, with populations of between 38,000 and 40,000. Bathurst is 200 kilometres from Sydney and Orange a further 60 kilometres away. Bathurst is Australia's oldest inland settlement. Well known for its connections to motor racing, it has a campus of Charles Sturt University (CSU) and a strong manufacturing and agricultural base (Bathurst Regional Council 2012). Orange too has a CSU campus and a recently opened hospital. Both towns are surrounded by rural areas characterised by vineyards and apple orchards. Both councils have marketed, and planned for further in-migration, through participation in the Evocities initiative, a name coined to emphasise that they were centres of Energy, Vision and Opportunity, and promoting them as regional 'cities', with similar services to capital cities, but with greater lifestyle attributes and affordability.

Surveys undertaken in 2011 demonstrate the distinctive characteristics of migrants to Bathurst and Orange, in terms of their age, origin and motivations for migration. Over 70 percent of a large sample (n= 1025) were migrants to both towns at some point in their lives (74 percent in the Bathurst cohort, and 69 percent

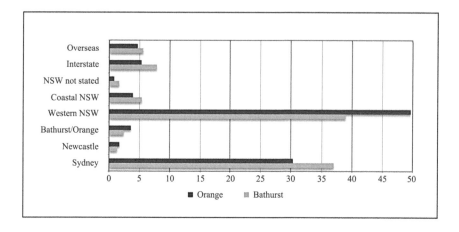

Figure 5.1   Migrants' Location of Origin (%)
            (Bathurst n = 373, Orange n = 360)

in the Orange cohort). People 65 years and over were overrepresented, both in the total sample (37.5 percent of respondents) and among the migrants (37 percent of migrants). While Sydney was a substantial source of migrants, most had moved from other towns in western NSW (Figure 5.1). At least half the migrants therefore were already familiar with living in rural and regional Australia. For this cohort, tree change emigration was not strictly about migration or flight from metropolitan centres, but was also concerned with perpetuating a regional/rural lifestyle. Precisely the same pattern recurs in the smaller NSW towns of Oberon and Glen Innes where a substantial part of migration was from other parts of regional NSW rather than from Sydney (Connell and McManus 2011). Remarkably the same pattern of migration into Bathurst and Orange existed some 35 years ago, when 29 percent of migrants to Orange and 36 percent of migrants to Bathurst originated from Sydney, but 44 percent in Orange and 37 percent in Bathurst originated came from elsewhere in regional NSW (Burnley 1980). For at least four decades there has been a steady flow of migrants from smaller rural and regional places into the larger towns.

By the early 1970s not only was the population of many inland country towns declining but regional populations were becoming concentrated in the larger towns, such as Horsham and Warrnambool in Victoria and Tamworth and Armidale in New South Wales (Holmes and Pullinger 1973, Burnley 1980). Some of that population growth resulted from migration from smaller nearby declining towns and rural regions, hence the growing centres became known as 'sponge cities'.

Services, especially government offices, and regional populations have become increasingly concentrated in larger towns such as Orange and Bathurst in NSW and Toowoomba and Rockhampton in Queensland. With greater access to cars, and thus long distance commuting, recreation and shopping, urban growth came at the expense of smaller towns. The towns most at risk of losing population were in the catchment areas of the sponges (Maher and Stimson 1994) rather than being particularly remote. Many small towns have become more like ghost towns, with services and populations gradually draining away (Forth and Howell 2002). However migration into Dubbo and Tamworth, two inland cities of NSW often seen as 'sponge cities', has been more complex with the population 'soaked up' from surrounding settlements only a relatively minor part of their growth. Such large towns, both with populations over 35,000, primarily drew new migrants from beyond their immediate hinterlands, from Sydney and outside the state (Argent *et al.* 2008). The 'sponge city' phenomenon is only partially true, though it is incontrovertible that larger towns are playing some part in sucking the life out of smaller towns, and that smaller towns have continued to decline fastest as their residents moved on or died. It is useful therefore to explore in more detail population changes in regional areas, such as Orange and Bathurst.

A third of migrants moved for reasons related to employment and business development (36 percent in Orange, and 32 percent in Bathurst) (Figure 5.2). Other significant influences on migration included: 'a better place to raise a family' (12

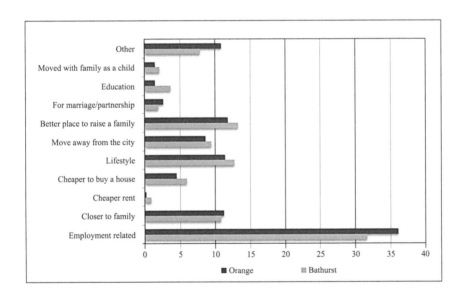

Figure 5.2   Motivation(s) for Migration (%)*

* Respondents could specify more than one option

   (Bathurst n = 575, Orange n = 536)

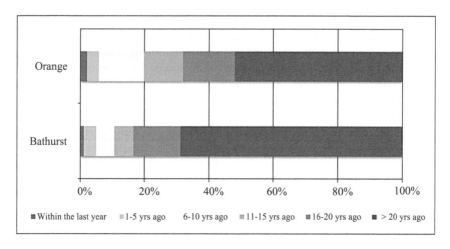

Figure 5.3   **Timing of Migration for Those who Moved for 'Employment Related Reasons'**

percent and 13 percent of migrants to Orange and Bathurst respectively), 'closer to family' and for 'lifestyle' reasons (11 percent of migrants to both towns). Employment underpinned most migration, whether through promotion, a new job or job transfer. The vast majority migrated when of working age or younger. Indeed, 69 percent of Bathurst in-migrants and 52 percent of Orange in-migrants moved for employment (Figure 5.3). Neither place was therefore characterised by retirement migration, further indicating the limited significance of lifestyle reasons in these migration streams, and the limited attractions that regional Australia holds for ageing populations (see Chapter 3).

Physical and Material Changes

Changes following migration can be examined through two different lenses: physical and material changes and the intangible or ambient changes. Physical and material changes include tangible changes in the provision of services and infrastructure – new or changed shopfronts and businesses – and changes to the physical landscape through residential development. Ambient and intangible changes include how locals have perceived their towns changing, and becoming less 'rural'. Locally-born Bathurst and Orange residents recognised that the provision of services and facilities had increased because of migration (Figures 5.4 and 5.5). Most viewed this as a sign of progress.

> There are a lot more activities to do in Orange than there was ten, 15 years ago. You can go out to the gardens [or the] wineries and take a tour of the mines.

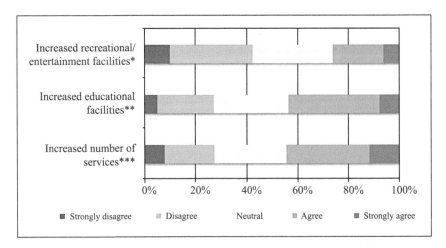

Figure 5.4　　Orange Locals' Opinions about the Changes in the Provision of Services and Facilities

*　　　Entertainment and recreational facilities included: cinemas, pools and theatres (n = 140)

**　　Educational facilities included: schools, universities and TAFEs (n = 144)

***　Services included: banks, ATMs, doctors, hospitals, childcare, government departments (n = 143)

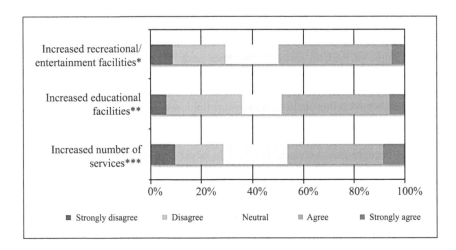

Figure 5.5　　Bathurst Locals' Opinions about the Changes in the Provision of Services and Facilities

*　　　Entertainment and recreational facilities included: cinemas, pools and theatres (n = 116)

**　　Educational facilities included: schools, universities and TAFEs (n = 115)

***　Services included: banks, ATMs, doctors, hospitals, childcare, government departments (n = 116)

It will cease to be the Bathurst that I grew up in but that's not a bad thing. Gaining a bigger community. A better place for our children to grow up in. A better future for our children's children.

A quarter of residents in both towns gave a more neutral response to these statements about service provision, with several suggesting that migrants brought a 'city mindset': the assumption that abundant services would exist on demand. As two Bathurst respondents contended:

We have never had any problem with medical services or any other [service] ... I get a little irritated by people who say "we should have exactly the same as the cities have. We should have the same sort of services they have in Sydney". Well you should have stayed in Sydney.

People [in-migrants] are less ... understanding of the rural community ... when services play up they're in trouble, they're not able to make do. They don't have the same resilience and self-reliance I guess that the rural people have traditionally had.

Such statements accord with the assertion that migrants' 'lifestyle choices ... remain mediated by their habitus and framed by their levels of social capital' (Benson and O'Reilly 2009: 618), as earlier preferences continue to influence life in their new destination, and become a source of disappointment, above all when people come from metropolitan centres. One migrant from Sydney stated: 'shops in Bathurst do not compare to those in Sydney, in range or service. [There is a] limited range of medical services'. Another said: '[I] did miss access to Sydney theatres and music venues. I have overcome this over the years, but still travel to Sydney for special occasions'. Thus 'certain individuals may live with a constant lack of fit between their habitus and social field, a permanent state of disruption of their social position' (Benson and O'Reilly 2009: 617, Ragusa 2011b). Perhaps unsurprisingly one Bathurst local contended that, while changes were beneficial and had brought new services, they might never live up to the expectations of people from Sydney: 'they're used to different lifestyles and different standards'. No place can ever live up to all expectations, especially for those who had come for employment and then sought services.

While new service provision changed the physical appearance of town centres, land use changes on the perimeter were identified as problematic. As populations grew, changing land use marked the transition from a productive agricultural sphere towards a multifunctional countryside, where 'market-driven urban interests' and amenity-oriented uses were (re)shaping rural space (Holmes 2006). Integral to land use change in Bathurst and Orange was the promotion of the rural and regional for its perceived aesthetic value and its associations with the lifestyle attributes of the rural idyll (Argent 2010, Holmes 2006). However rural land conversion accompanied the development of new residential housing areas and the consequent loss of agricultural land. While site specific, that was also 'influenced by broader scale processes and events', namely deregulation and the reduction of protectionist policies (Argent 2010). In Orange, land subdivided for residential development replaced orchards and dairy farms, changes that were strongly felt at a local level. Historically, Bathurst and Orange's economies were linked to agriculture, creating and contributing to a 'rural' imagery, where farms and other agricultural endeavours prevailed over urban images of tightly-packed residential subdivisions. Greater densities and urban sprawl were unwelcome. One local commented on the aesthetics of new housing estates in Orange saying: 'I think it's pretty crowded, it's a bit like Lego land, I would like to see more rural'. An Orange City Council document, *Orange 2020*, argued that 'fertile agricultural land should be protected from development and urban sprawl with no further expansion beyond the city limits' (Orange City Council 2007: 6). Yet, the document also outlined a strategy for continued growth using the slogan 'the best balance between city and country' (Orange City Council 2007: 11), anticipating that there should be 'fewer restrictions on rural and residential developments' (Orange City Council 2007: 6).

Participation in the Evocities venture has seen both towns (re)position their rurality by endorsing country attributes alongside city-style living and amenities. The crux of this marketing campaign, as for the 26 percent of respondents who viewed these locations as providing a lifestyle change and a better place to raise a family (Figure 5.2), was the attractiveness of the rural location. Through place branding, the country is viewed both as a 'traditional ... and discursive site for family, community and well-being' and 'a commercial selling point and a commodified imaginary' (Gorman-Murray 2008: 1). In Bathurst, the rural lifestyle was promoted in relation to housing affordability, a country lifestyle and family living. In Orange, the existing tourist infrastructure, including the 'regular seasonal food, wine and arts festival[s]' had already, to the Council, positioned it as the 'perfect city to bring up a family' (Evocities 2012). Central to the imaginings of both locations are constructions of rurality; as two residents suggested by having a 'slower lifestyle and pace, community, quietness, [and a] lack of pollution' and a 'friendly country town character with excellent services'. These imaginings of place form part of the drawcards in the promotion of these (and many other) tree change locations as intimately intertwined with the idylls, and ideals, of rurality. Yet they are fragile and often misconceived, with newcomers themselves spoiling the tranquillity, despite their complaints of farm machinery being noisy (Connell and McManus 2011). Migration has constantly changed both the nature of place and how locals and new migrants experience such places.

Opinions about rural change centred on two interconnected themes: the loss of agricultural land, and the expansion of new residential land. The majority of Bathurst and Orange's migrants and locals specified that too many new housing developments jeopardised the regional lifestyle and the character of the town. However more migrants agreed (57 percent of Orange and 48 percent of Bathurst migrants) with the statement 'I think that new residential land releases (on former farmland) jeopardise the regional character', than did locals (48 percent of Orange and 47 percent of Bathurst locals agreed or strongly agreed). Migrants were more likely to emphasise the importance of maintaining a distinctive regional/rural landscape than locals:

> We came to the Bathurst region because in the city you were just getting built out and built out and built out … then when we got here we found out they were doing the same thing here.

> I moved out of Sydney with that gentrification thing going on all around me, and now it's happening again. Here we go, it will be cut down every tree, lawn, six iceberg roses, whack down the two bedroom old house, and then the next thing you know there's people coming round putting in four units.

In Orange three quarters of both locals and migrants argued that agricultural land should be protected, yet two thirds also thought that there should be fewer development restrictions.

> Fifty years ago, we were producing prime lambs in this part of the world and there were five lambs to the acre and the farming was good solid quality …. 500 to 5000 acre [farms], today you can drive from here to Bathurst to Blayney and … what will you find? Ten acre, five acre [farm]s, [with a] broken down old cow … a broken down old house … a few dogs and a cat, that's what our prime farms are being turned into. They're being destroyed, and one day we will wake up, well I won't be alive, but somebody will wake up and say it's covered in concrete.

> There's lots of visual amenity, like it's really nice to look out and there's some green paddocks over there, I see all those suckers buying five acre blocks out of town, you know, like their little bit of rural whatever it is, and I know that in a couple of years they'll all be filled up by streets and things.

Concern over the loss of agricultural land, and thus of rural landscape, emphasises the role that a rural outlook plays in many decisions to live in Orange. Hobby farms have been widely resented elsewhere, as obvious forms of 'creative destruction', as in the Southern Highlands and the evolving 'postmodern' agricultural economy of Bridgetown, Western Australia (Tonts and Greive 2002, Argent *et al.* 2011, Drozdzewski *et al.* 2011). In north Queensland such divisions had even threatened the viability of the rural economy (Bohnet and Moore 2011).

Along with visible changes many people perceived intangible yet important changes, which were overwhelmingly viewed unfavourably through sentiments imbued with a sense of loss, articulated as a decline in the bucolic ambience of place. Common sentiments from both locals and migrants concerned the towns becoming too like Sydney or not 'country' enough:

> It's become ... a bit more modern, more city-fied [*sic*], and less country town.
> I live out in the bush and I think, there's no way you'd call the city of Orange
> bush, I could be in a suburb of Sydney.

> Bathurst is becoming too big and is losing its country feeling.

> What really concerns me [is that] they advertise Bathurst as the place to be, they
> advertise the rural beauty and it's atmosphere and ... they are destroying it as
> fast as they can go.

Such changes were routinely attributed to people coming from the city, despite
most migrants coming from other parts of regional New South Wales. Locals were
more likely to express concern over changes with some reference to agricultural
character. The rural character of the towns' agricultural shows, the movement
away of sale yards and increasing urbanity all figured in these sentiments. As one
Bathurst respondent commented:

> There'll be some people here that have never been on a farm. There will be a lot
> of people living in Bathurst that will have never been on a farm. Twenty years
> ago that might not have been the case. But it certainly is now.

The loss of rural character prompted one Orange local to declare that, 'they
[migrants from the city] have no concept whatsoever of what a farm is', despite
living in regional Australia. For locals, who had witnessed changes to agricultural
character, benefits from continued migration were weighed against the loss of a
much-cherished rural lifestyle and landscape, and steady urban sprawl. Despite
both Bathurst and Orange being large regional centres, locals' conceptions and
memories of place were embedded in a particular rural imaginary both agricultural
and Arcadian.

Small Places

Despite their perceived need to market themselves, Bathurst and Orange have
proved relatively attractive and growing destinations, because they have viable
regional economies – partly by being service centres, and within 300 kilometres of
Sydney. Rural areas that have become primary tree change destinations are close to
large metropolitan areas, with attractive scenic environments, a circumstance that
steadily declines inland, often associated with wine-growing regions. Many such
'hotspots' are actually coastal – the beneficiaries of sea change – but others such
as the Barossa Valley, Mudgee, Bowral, Castlemaine and Daylesford are inland.
Image is important and even large inland cities such as Wagga Wagga have joined
Evocities and positioned themselves as tree change destinations in advertising and
marketing directed at wealthy city dwellers (Ragusa 2011a). In some respects that

matches the aspirations of tree changers, most of whom – as many as 68 percent in one NSW study – preferred to move to towns and not to rural properties (Ragusa 2011a), indicating that the rural environment – as one facet of lifestyle – was not an important criterion. New migrants were not interested in seeing a farm, but a rural vista was welcome and a seemingly decisive element of migration.

Throughout Australia a pattern recurs of relatively stable regional growth centres, not too far from the coast, population decline further inland and the struggle of smaller places to thrive and prosper. Indeed small inland centres have found the task of retaining people so great that marketing is pointless. Councils in western NSW do not attend Country Week, while even larger towns that are far from metropolitan centre, such as Moree, have only attended intermittently, confronted with the reluctance of metropolitan populations to move beyond a certain comfort zone. Thus in NSW Oberon has attracted more new migrants than Glen Innes simply because it is much closer to Sydney (Connell and McManus 2011). The ability to combine some degree of rural living with maintained urban connections has been crucial. Consequently, like Oberon but unlike inland regions, the majority (59 percent) of migrants to the Southern Highlands, about 120 kilometres from Sydney and no more that 30 kilometres inland, had come from Sydney and cited 'lifestyle' and 'to move away from the city' as key explanations. Most of the other migrants to the Highlands had come short distances from elsewhere in NSW, and especially from the nearby Illawarra region. As many as 77 percent of the new migrants chose the Southern Highlands because of its proximity to Sydney, and the ability to combine a more relaxed country lifestyle with access to friends, family, employment, health facilities and the airport. In places like Katoomba, Castlemaine and Kangaroo Valley, all not much more than 100 kilometres from capital cities, tree changing merges into 'extreme commuting' (Butt 2011). In a sense each such place has increasingly become a metropolitan village rather than a place apart.

New migrants in the Southern Highlands as elsewhere said: 'We wanted our children to have a childhood more like ours – more freedom and sense of community that is not achievable in Sydney' and 'After 10 years in Sydney, we longed to get back to a country town: less busy, less activity' (Drozdzewski *et al.* 2011). As in every other tree change location, migrants often had some previous connection to the region, or to rural Australia: they were familiar with particular places or still had friends and relatives there and were more easily able and willing to make the move.

Tree change migration has not always been successful. New more positive lifestyles were not necessarily achieved and lack of services posed problems for many. Even in the 1970s, as new migrants moved to Bathurst and Orange some, women especially, experienced problems – notably finding suitable accommodation and employment – ironically the attractions of Evocities that were later marketed. Lack of services, such as pre-school facilities, was exacerbated by the lack of social support, with close relatives often elsewhere (Burnley 1980). Over time and in rather smaller and more remote places such problems recurred (Connell and McManus 2011). While small towns offered relative value for money in terms of housing the cost of living was often higher than anticipated, even for fruit and vegetables, while shops closed on Sundays (Ragusa 2011a, Connell and McManus 2011). Such problems were invariably greatest for those who had moved from state capitals rather than from other regional towns. Bucolic landscapes, peace, space and security for children and lack of pollution could not necessarily compensate for reduced access to amenities and services. However in the Southern Highlands, close to two large cities, many new residents were willing to forgo the lack of some services because of the relative ease of access to Sydney. The umbilical cord that tied tree changers to their place of origin proved crucial.

Simply fitting in to a new place could be difficult. Tree changers were not locals and however hard some tried to fit in they were not always welcomed by locals who could resist social change, however beneficial that might seem to be in terms of new services and amenities. 'You have to be born and bred here to be a local' and even provide evidence of relatives in the cemetery. Newcomers were 'blow-ins' who might just as easily blow out again (Drozdzewski *et al.* 2011). As one Bathurst resident contended: 'It has been quite challenging having not been born in Bathurst to fit in to being a local. There is a definite line between being born here and moving here'. Newcomers craved community, and were anxious to find it, but perceived transients were unwelcome and permanency took time to establish. In many places new migrants eventually formed communities and established social networks mainly with other tree changers (Ragusa 2011b). That was rather easier in places like the Southern Highlands where so many people were tree changers and had come from Sydney. Unsurprisingly most successful tree changes have occurred in places closest to the sources of migrants, notably the Southern Highlands, where migrants could retain some connections with the places they had come from, and could experience rural, urban and metropolitan.

Conclusion

Australian studies have demonstrated the diversity of tree change migration, supporting previous contentions that tree change migrants are difficult to characterise easily into distinct groups (Connell and McManus 2011, Ragusa 2011b, Halfacree and Rivera 2012). Popular representations of tree change migrants 'frozen into media-friendly stable categories' assume that tree change migrants come only from the city and are searching for lifestyle change (Halfacree 2008: 483, Ragusa 2011a). Yet these characteristics did not typify most migrants to country towns. They came from elsewhere in the state as much as from state capitals, were engaged in a myriad of occupations, were of various ages, occupations and countries of birth, and above all had different motivations. Neither lifestyle nor retirement was particularly important compared with employment and housing affordability (Ragusa 2011b). Except in places like the Southern Highlands, close to the city, it has become increasingly evident that the popular perception of tree change migrants moving for lifestyle reasons from metropolitan cities is wrong; the majority of 'new' migrants in the country are not escapees from urban life, but people who have moved within rural regions, perhaps from smaller towns, but as often because they were transferred rather than because they sought out particular idyllic spots. Others were returning to places they had once been familiar with or where they had relatives and friends (Connell and McManus 2011). Only in a handful of places, not far from metropolitan areas, again like the Southern Highlands and Castlemaine, have most newcomers migrated from these cities – in some respects becoming a detached and distant part of suburbia. Moreover it is probably those new rural suburbanites and 'urban refugees', who have gone no further than such places as the Blue Mountains, the Barossa and Hunter Valleys, within an hour or so commuting from cities, who have been able to make the easiest adjustment to 'rural' areas. The most successful tree changers are either still part of perimetropolitan areas, experiencing the best of both worlds, or those who have moved to larger towns like Bathurst and Orange, where a range of services exist. One way or another they had moved out of town but not out of touch (Connell and McManus 2011). Otherwise, and at least across the 'sandstone divide' of so much of eastern Australia, it has even been said that urban residents exhibit 'a grand prejudice arising from our collective ignorance' that prevents tree change migration (Iyer 2012: 1). Repopulating remote regions and small places is a massive task.

Ironically migrants from metropolitan centres were most likely to prefer the rural ambience of regional towns, while longstanding local residents – anxious to retain the services they have and even improve upon them – were more likely to welcome population growth, despite the pressures this sometimes entailed. Hence even large towns like Bathurst and Orange have participated in the Evocities programme, and smaller centres been part of Country Week and other forms of place marketing. The importance of maintaining rural imagery has been universally popular, most obviously in larger towns like Bathurst and Orange, where physical change was most visible. No-one welcomed higher density living, and locals particularly mourned the loss of the agricultural character of their towns. Places are contested and precisely these aspects of place are at most risk from tree change and the continued promotion of migration, where that promotion focuses on access to services and 'modern' amenities. Rural imagery has remained important in how places market themselves as tree change locations, while mixing the availability of city experiences with country lifestyles. Notions of 'lifestyle' and 'community' are never far away. As far as possible place marketing is all things to all people.

Tree change has restructured regional and rural landscapes. What has emerged is a more 'hybrid country', one no longer solely related to agriculture or necessarily to a narrative of rural decline, but where the idealised and actual imaginary of rural life continue to play an important part in how residents articulate and seek to maintain regional/rural lifestyles, but at the same time seek out urban services: a new hybrid 'micropolitan' context (Connell and McManus 2011). Increasingly therefore regional populations have become concentrated in medium sized 'sponge' towns and small cities, like the Evocities, and marketing emphasises the virtues of the service provision that such towns can offer and that smaller places cannot. In turn this has meant that certain places are particularly favoured, notably those towns and cities that are within a few hundred kilometres or a few hours' drive of metropolises, at the expense of those much further inland. Ironically, but perhaps appropriately, this has meant that tree change migration has gone to the borders of a more coastal Australia where trees remain common and stopped where wide open plains take over.

References

Argent, N., 2008, 'Perceived Density, Social Interaction and Morale in New South Wales Rural Communities', *Journal of Rural Studies*, 24: 245-61.

Argent, N., 2010, 'Regions and Communities Dividing? Australian Rural Development in a Multifunctional Context', in D. Winchell, D. Ramsey, R. Koster and G. Robinson (eds), *Geographical Perspectives on Sustainable Rural Change*, Brandon: Rural Development Institute and Brandon University, 330-347.

Argent, N., Rolley, F. and Walmsley, D.J., 2008, 'The Sponge City Hypothesis: Does It Hold Water?', *Australian Geographer*, 39: 109-30.

Argent, N., Tonts, M., Jones, R. and Holmes, J., 2011, 'Amenity-Led Migration in Rural Australia: A New Driver of Local Demographic and Environmental Change', in G. Luck, D. Race and R. Black (eds), *Demographic Change in Australia's Rural Landscapes*, Dordrecht: Springer, 23-44.

Bathurst-Orange Development Corporation, 1978, *Room to Grow*, Bathurst, New South Wales.

Bathurst Regional Council, 2012, Live in Bathurst Available at http://www.bathurstregion.com.au/live/ [accessed 16 May, 2012].

Beer, A., Maude, A. and Pritchard, B., 2003, *Developing Regional Australia*, Sydney: UNSW Press.

Bell, M.J., 1992, *Internal Migration in Australia 1981-1986*, Canberra: Australian Government Publishing Service.

Benson, M. and O'Reilly, K., 2009, 'Migration and the Search for a Better Way of Life: A Critical Exploration of Lifestyle Migration', *Sociological Review*, 57: 608-625.

Bohnet, I.C. and Moore, N., 2011, 'Sea- and Tree-Change Phenomena in Far North Queensland, Australia: Impacts of Land Use Change and Mitigation Potential', in G. Luck, D. Race and R. Black (eds), *Demographic Change in Australia's Rural Landscapes*, Dordrecht: Springer, 45-69.

Burnley, I.H., 1980, 'Migration to Larger Country Towns: The Bathurst-Orange Case', in I. Burnley, R. Pryor and D. Rowland (eds), *Mobility and Community Change in Australia*, Brisbane: University of Queensland Press, 38-47.

Burnley, I.H., 1988, 'Population Turnaround and the Peopling of the Countryside? Migration from Sydney to the Country Districts of New South Wales', *Australian Geographer*, 19: 263-83.

Burnley, I.H., Pryor, R.J. and Rowland, D.T., 1980, *Mobility and Community Change in Australia*, Brisbane: Queensland University Press.

Burnley, I.H. and Murphy, P.A., 2002, 'Change, Continuity or Cycles: The Population Turnaround in NSW', *Journal of Population Research*, 19: 137-154.

Burnley, I.H. and Murphy, P., 2004, *Sea Change. Movement From Metropolitan to Arcadian Australia*, Sydney: UNSW Press.

Butt, A., 2011, 'The Country Town and the City Network: The Expanding Commuter Field of Melbourne', in J. Martin and T. Budge (eds), *The Sustainability of Australia's Country Towns: Renewal, Renaissance, Resilience*, Ballarat: Victorian Universities Regional Research Network Press, 23-57.

Collits, P., 2000, *Small Town Decline and Survival: Trends, Success Factors and Policy Issues*, Bendigo: Regional Institute Limited, 1-49.

Connell, J. and McManus, P., 2011, *Rural Revival? Place Marketing, Tree Change and Regional Migration in Australia*, Farnham: Ashgate Publishing.

Costello, L., 2007, 'Going Bush: The Implications of Urban-rural Migration', *Geographical Research*, 45: 85-94.

Drozdzewski, D., Shaw, W. and Godfrey, N., 2011, *Treechange Migration to the NSW Southern Highlands: Motivations and Outcomes of Urban to Rural Migration*, Report to NSW Department of Regional Development and Local Government, Sydney.

Evocities, 2012, *A Big Move can be a Challenge.* Available at: http://www.evocities.com.au/ [accessed 1 June 2012].

Forth, G. and Howell, K., 2002, 'Don't Cry for Me Upper Wombat: The Realities of Regional/Small Town Decline in Non-coastal Australia', *Sustaining Regions*, 2: 4-11.

Gorman-Murray, A., 2008. Country, *M/C Journal*, 11(5). Available at: http://journal.media-culture.org.au/index.php/mcjournal/article/viewArticle/102.

Gray, I. and Lawrence, G., 2001, *A Future for Regional Australia: Escaping Global Misfortune*, Cambridge: Cambridge University Press.

Gurran, N., 2008, 'The Turning Tide: Amenity Migration in Coastal Australia', *International Planning Studies*, 13: 391-414.

Gurran, N., Squires, C. and Blakely, E., 2005, *Meeting the Sea Change Challenge*, Report for the National Sea Change Taskforce, Sydney.

Gurran, N. and Blakely, E., 2007, 'Suffer a Sea Change? Contrasting Perspectives Towards Urban Policy and Migration in Coastal Australia', *Australian Geographer*, 38: 113-132.

Halfacree, K., 2008, 'To Revitalise Counterurbanisation Research? Recognising an Intentional and Fuller Picture', *Population, Space and Place*, 14: 479-495.

Halfacree, K. and Rivera, M.J., 2012, 'Moving to the Countryside ... and Staying: Lives beyond Representations', *Sociologia Ruralis*, 52: 92-114.

Holmes, J., 2006, 'Impulses Towards a Multifunctional Transition in Rural Australia: Gaps in the Research Agenda', *Journal of Rural Studies*, 22: 142-160.

Holmes, J. and Pullinger, B., 1973, 'Tamworth: An Emerging Rural Capital?', *Australian Geographer*, 12: 207-225.

Haxton, N., 2005, 'From Seachange to Treechange, Seachange becomes Treechange'. Australian Broadcasting Corporation transcripts (17 August).

Hugo, G., 1994, 'The Turnaround in Australia: Some Observations from the 1991 Census', *Australian Geographical Studies*, 25: 1-17.

Iyer, P., 2012, 'Crossing the Great Lifestyle Divide', *The Weekend Australian*, 25 August, 1.

Little, J. and Austin, P., 1996, 'Women and the Rural Idyll', *Journal of Rural Studies*, 12: 101-111.

Maher, C. and Stimson, R., 1994, *Regional Population Growth in Australia: Nature, Impact and Implications*, Canberra: Australian Government Publishing Service.

Nicholls, S., 2011, 'Public Not Moved by Regional Relocation', *The Sydney Morning Herald*, 11 October.

Orange City Council, 2007, *Expressed Community Views: Orange 2020 Community*, Orange.

Ragusa, A., 2011a, 'Seeking Trees or Escaping Traffic? Socio-cultural Factors and "Tree-change" Migration in Australia', in G. Luck, D. Race and R. Black (eds), *Demographic Change in Australia's Rural Landscapes*, Dordrecht: Springer, 71-100.

Ragusa, A.T., 2011b, 'Changing Towns, Changing Culture: Examining Tree Changers' Perceptions of Community in Australian Country Towns and Places', in J. Martin and T. Budge (eds), *The Sustainability of Australia's Country Towns: Renewal, Renaissance, Resilience*, Ballarat: Victorian Universities Regional Research Network Press, 81-103.

Salt, B., 2001, *The Big Shift*, Melbourne: Hardie Grant Books.

Smailes, P., 2002, 'From Rural Dilution to Multifunctional Countryside: some Pointers to the Future from South Australia', *Australian Geographer*, 33: 79-95.

Tonts, M. and Grieve, S., 2002, 'Commodification and Creative Destruction in the Australian Rural Landscape: The Case of Bridgetown, Western Australia', *Australian Geographical Studies*, 40: 58-70.

Walmsley, D.J., Epps, W.R. and Duncan, C.J., 1998, 'Migration to the New South Wales North Coast 1986-1991: Lifestyle Motivated Counterurbanisation', *Geoforum*, 29: 105-118.

Ware, V.A., Gronda, H. and Vitis, L., 2010, *Addressing Locational Disadvantage Effectively*, Sydney: Housing NSW.

Winchester, H.P.M. and Rofe, M.W., 2005, 'Christmas in the "Valley of Praise": Intersections of the Rural Idyll, Heritage and Community in Lobethal, South Australia', *Journal of Rural Studies*, 21: 265-279.

Chapter 6
Difference or Equality? Settlement Dilemmas on the Indigenous Estate

John Taylor

Most of what constitutes rural and regional Australia is now subject to some form of indigenous land tenure or co-management regime, or is under legal claim as such. This is a surprising revelation to many because indigenous peoples and their lands tend to be absent from popular imaginings of the bush and from current policy and academic discourse on rural and regional issues (NIRRA 2012). In turn, this may have been because the transition underway in rural and regional Australia, where indigenous values are contesting the former dominance of monocultural pastoral and agricultural production. This has occurred gradually and in piecemeal fashion since the 1970s, though it has gathered pace in recent years (Holmes 2010). Incrementalism here reflects the varied manner in which much of Commonwealth, State and Territory legislation has emerged over the past few decades to give effect to indigenous land rights with the outcomes still unfolding. In total, eight pieces of Commonwealth legislation and 28 across the States and Territories have been enacted over the past 40 years that have had varying impact on indigenous access to land and resources (AHRC 2011: 103-4, Davies 2003, Tehan and Godden 2012). This chapter examines how the accumulation of land transfers to indigenous title has served to create a new geography of rural and regional Australia and how this, in turn, has stimulated debate about the viability and future direction of indigenous modes of rural occupance.

The amalgam of land holdings produced by these myriad of legal processes has been referred to as an 'indigenous estate' (Pollack 2001, Altman *et al.* 2007) a somewhat loose term given the different bundles of rights that it embraces but one, nonetheless, that demarcates a structural indigenous geography not dissimilar in conceptualisation to an earlier notion of an 'indigenous domain' as spaces where indigenous peoples and their institutions predominate (von Sturmer 1984, Trigger 1986). While fundamentally determined by jurisprudence, the notion of the indigenous estate embraces much more than a demarcation of land tenure arrangements: it is acknowledgement of an on-going shift in political economy whereby the return of country to original owners provides a foundation for indigenous livelihoods and a negotiation over the sharing of land and resources. This is geographically manifest as an 'indigenous occupance mode'

(Holmes 2010), and while it is sustained by what Altman (2001) describes as a 'hybrid economy' with customary or non-market activities intersecting to varying degrees with both market and state sectors, it is characterised by 'high welfare dependency and associated severe problems of dysfunctionality' (Holmes 2010: 353). It becomes, therefore, an analytic category, a testing ground for what Austin-Broos (2011) labels the 'politics of difference and equality' and others see as an experiment in 'interculturality' (Rowse 2012). The equality that Austin-Broos alludes to here refers to attempts by the state and many indigenous people to overcome pathologically poor social and economic outcomes that are commonplace on the indigenous estate (Pearson 2000); the difference refers to those who view these attempts as a neoliberal undermining of cultural choice and opportunities for alternative development (Altman 2012). Either way, both positions struggle to address the legacy of a colonial system that has left indigenous people on country 'with modest but sufficient means to disengage from market society ... while not equipping them to avoid the panoply of ill-health and distress that comes with marginalisation' (Austin-Broos 2011: 46).

One element of this dilemma that informs the present chapter concerns the unique demographic response to the emergence of an indigenous estate and the degree to which this is implicated in a tension between the desire of many indigenous people to continue to reside on traditional country, on the one hand, and the centralising tendencies of state policies and the market, on the other. For the most part, across rural and regional Australia, market forces have dominated and recent demographic history has been one of chronic population decline and ageing due to selective out-migration and a contraction of settlement as a consequence of economic restructuring and adjustment (McKenzie 1994, McGuirk and Argent 2011). On the other hand, in those areas now designated as indigenous lands, the trajectory has been quite different – population growth has been considerable due to high natural increase, there is considerable mobility but far less migration, and there has been a proliferation of new settlement as 'outstations' or 'homeland centres' although, interestingly, these have emerged alongside embryonic urbanisation. An emerging issue, then, over a growing expanse of rural Australia, is less to do with managing decline and transformation and much more to do with accommodating (often literally) an escalation of indigenous occupance and providing adequate services and life opportunities for populations that to date have had little experience of either.

Recognition of land rights along with universal access to income support has therefore had profound implications for the geography of indigenous Australia by providing a land base for proprietary use and a means to sustain attachment to traditional country for a population that was previously expected to have migrated to urban centres or industrial sites owing to the lack of economic opportunity in some of the remotest parts of the continent (Rowley 1971). But because socioeconomic outcomes for indigenous residents are persistently low, and because the state remains the dominant economic provider via program spending and income transfers, the emerging mode of occupance is inherently vulnerable to external influence and a key policy debate has arisen regarding its viability and future direction.

The Indigenous Estate: An Emerging Spatial Construct

The return of land to indigenous interests began with passage of the *Aboriginal Lands Trust Act 1966* in South Australia, but the Commonwealth's *Aboriginal Land Rights (Northern Territory) Act 1976* was the more significant legislation as it provided an argument for land rights based on the recognition of ritual and economic use of the land and an established system of indigenous law. To date this has led to the transfer of almost 50 percent of the Northern Territory, with around 600,000 km² converted to non-transferable freehold tenure and held by traditional owners as a *de facto* property right. Similar legislation was subsequently passed in South Australia with the *Anangu Pitjantjatjara Yankunytjatjara Land Rights Act 1981* and the *Maralinga Tjarutja Land Rights Act 1984.* By the 1990s, all other states and territories had established statutory arrangements for the granting of a variety of forms of indigenous land title mostly by conferring access to vacant Crown lands.

In terms of spatial impact, the most significant event has been the recognition of native title by the High Court in *Mabo v Queensland (No. 2)* (1992) 175CLR1 *(Mabo)* and the subsequent title management provisions of the *Native Title Act 1993 (Cth) (NTA)* as this has resulted in a proliferation of determinations of native title as well as registered claims for native title that now extend over vast areas of rural and regional Australia. While the *Mabo* judgement in Queensland conferred much weaker rights than the legislative regimes in the Northern Territory and South Australia (not least because native title can be extinguished), it nonetheless requires and recognises customary non-market property rights related to land and sea resources and ensures that indigenous interests have a (restricted) right to negotiate with other potential land users, or at least to be consulted, which opens up important opportunities to leverage commercial and other benefits (Altman 2012, O'Faircheallaigh 2012, Tehan and Godden 2012). In this regard, an important 1998 amendment to the NTA introduced the option of establishing an Indigenous Land Use Agreement (ILUA) as a voluntary agreement between a native title group and others about the use and management of land and water that allows parties to negotiate pragmatic arrangements for co-existence. These can be established over areas where native title has, or has not yet, been determined and they provide an option for resolving native title issues through timely negotiation rather than costly and time consuming litigation. Over the past decade they have provided for significant additions to the indigenous estate, especially in rural Queensland.

Between 1994 and 2011, a total of 134 determinations of native title were declared amounting to 1.2 million km² or approximately 15 percent of the Australian land mass. Almost one half (48 percent) of this area is declared as so-called 'exclusive possession' which confers a right to possess and occupy an area to the exclusion of all others and to control access to and use of land. This exclusive title is overwhelmingly found in Western Australia. Elsewhere, native title is predominantly 'non-exclusive' and these rights are waived. In some instances determinations cover urban areas, as in the Global Agreement between Yawuru people, the Shire of Broome and the government of Western Australia that formalises joint management of lands and waters in and around the town of Broome and provides native title holders with substantial land-holdings and investment opportunities. This agreement includes just two out of a total of 628 registered ILUAs around the country that now cover 17 percent of the land mass including areas that are not yet the subject of native title determinations. In addition, conservation initiatives under the National Reserve System (NRS) also provide for the declaration of Indigenous Protected Areas (IPAs). These are areas of Indigenous-owned land or sea where traditional owners have entered into an agreement with the Australian Government to promote biodiversity and cultural resource conservation (Baker *et al.* 2001, Langton *et al.* 2005). There are now 50 declared IPAs covering over 260,000 km² and accounting for almost one-quarter of the NRS. Finally, the Indigenous Land Corporation (ILC), a statutory body established in 1995 to invest returns from the Aboriginal and Torres Strait Islander Land Account, acquires and grants land to Indigenous corporations to provide access to and protection of cultural and environmental benefits and socioeconomic development (Altman and Pollack 2001). Since 1995, the ILC has acquired and returned to indigenous tenure a total of 238 rural properties in all States and Territories amounting to 62,000 km².

Elsewhere, there are comprehensive agreements that can encompass whole regions. In the south west of Western Australia, for example, the South West Aboriginal Land and Sea Council (SWALSC) is negotiating an out of court settlement with the State government regarding six native title claims over the metropolitan area of Perth and surrounding region involving the transfer to Aboriginal ownership of several hundred thousand hectares of land and joint management of national parks on behalf of the Noongar nation. In Victoria, the *Traditional Owner Settlement Act 2010 (Vic)* creates a framework for the State Government to make a series of agreements covering land, land use activities and natural resource management with specific entities that represent traditional owner groups where the land is unreserved public land or where the grant of a new 'Aboriginal title' can be made over public land. As in the SWALSC case, this is seen as providing an alternative to the resolution of native title under the *Native Title Act* and is intended to provide for more speedy hand back/lease back of land to an underlying form of Aboriginal title.

Clearly, there has been and remains substantial land transfer activity across rural and regional Australia but keeping track of these developments is difficult since there is no single national database that can indicate the quantum and location of land held under indigenous title or that is subject to some form

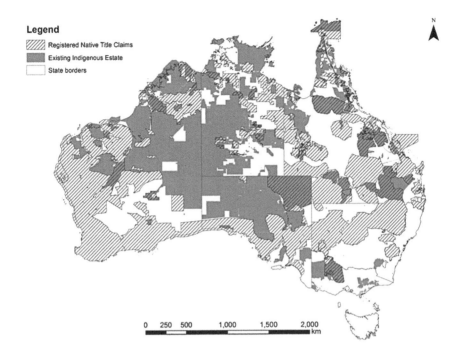

Figure 6.1   The Indigenous Estate and Registered Claims for Native Title, 2012

Source: NNTT 2012, Altman *et al.* 2007

of legal claim as such. The first attempt to develop a comprehensive national tally was the work of Pollack (2001) with a subsequent update by Altman, Buchanan and Larsen (2007). The Productivity Commission now provides an annual summary of indigenous-held lands in its annual *Overcoming Indigenous Disadvantage* report by recording additions to the indigenous estate via native title determinations and ILUAs (SCRGSP 2011). What these various efforts reveal is a spatial patchwork of indigenous rights to own land and/or to negotiate, to be consulted on, and to participate in matters and activities affecting the utilisation of land, seas and waters to an extent that can no longer be overlooked as an integral part of doing business in rural and regional Australia. Approximately 31 percent of the Australian continent is now subject to some form of indigenous land tenure while a further 45 percent has passed the registration test for native title determination. Figure 6.1 shows the distribution of these lands in two broad groupings: the first combines areas that have been returned to indigenous interests under land rights legislation, as a native title determination, as an ILUA or as an IPA (these categories often overlap); the second shows lands that have passed the government's formal registration test for native title.

Areas where some form of indigenous title has been declared are concentrated in remote parts of Australia and include large tracts of territory across central Australia and areas of northern Australia such as the Kimberley, Arnhem Land and Cape York. Elsewhere, indigenous tenure is patchy, especially across Queensland and New South Wales where numerous parcels of Aboriginal land allocated under the *Aboriginal Land Rights Act 1983 (NSW)* are too small to be represented at the scale shown here. Several IPAs in New South Wales, Victoria and Tasmania are also hard to display for the same reason. Land registered under claim for native title is even more extensive and distributed across regional areas in the south of Western Australia, South Australia and across central New South Wales with other major claims in the Pilbara region of Western Australia and much of southern Queensland. While there is no guarantee that native title will be declared for these areas, even the prospect of native title can stimulate the negotiation of alternate co-existence arrangements such as an ILUA and a framework is therefore created for the inclusion of native title parties in discussions over land use.

Aside from the vast area of land implicated, the significance of this new geography is reflected in the proportion of the indigenous population that it encapsulates. In 2011, 19 percent of the indigenous population lived on land rights lands, native title lands, ILUAs or IPAs while a further 26 percent lived in areas under claim for native title. These figures are approximations since it is often difficult to match cadastral boundaries to census geography, but the fact of indigenous people's interest in land is undeniable: in the 2008 National Aboriginal and Torres Strait Islander Social Survey, for example, two-thirds (62 percent) of the indigenous adult population indicated that they identified with a clan or language group and as many as three quarters (74 percent) recognised an area as their homeland or traditional country. Nationally, 25 percent indicated that they lived on their homelands/traditional country; a similar proportion was reported in regional areas but it rose to more than one half in very remote areas. Obviously, much lower rates were reported in major cities.

The Indigenous Occupance Mode

A tangible impact of this land transfer has been growth of the rural population and revitalisation of dispersed settlement. This contrasts with the more common experience in rural areas that has seen population decline and a contraction of small rural localities. A largely unrecognised prop to regional economies that might otherwise have contracted even further has been the public (and private) employment and investment that has derived from the delivery of services and infrastructure to provide for this expanding indigenous domain. While attempts to highlight this for particular regions have uncovered the substantial contribution to regional accounts of production on indigenous lands and expenditures on indigenous-related services (Altman and Kerins 2012, Gerritsen *et al*. 2010, Crough and Christophersen 1993) other studies have questioned the adequacy of public investment on equity grounds due to inappropriate distributive mechanisms and census undercounting of indigenous populations (Taylor and Stanley 2005, Taylor and Biddle 2010). To emphasise the contemporary significance of this it is interesting to reflect on what, in the 1960s was thought of the plight of rural indigenous peoples just prior to the emergence of land rights.

Conscious of a growing mood for change, the (then) Social Science Research Council of Australia sponsored a series of studies on 'Aborigines in Australian Society' that focused, among other things, on documenting the dire social and economic conditions that prevailed in rural reserve locations where indigenous people had become required to live. One of the more influential of these studies described the various mission and government settlements established for indigenous people at that time as instrumental in frustrating their urbanisation. These places were seen as 'holding institutions' serving to prevent the inevitable migration of indigenous people to towns and cities (Rowley 1971: 84). With the benefit of 50 years hindsight during which time indigenous people have been free from institutional and legislative constraints on their place of residence, this proposition is only partly valid. While migration of indigenous people from remote areas to regional areas and cities has undoubtedly occurred, and still does in a step-wise fashion (Biddle 2009), the overall net flow has been only marginally positive since the 1970s. Consequently, much of the dramatic growth in urban indigenous population that has been observed in recent decades simply reflects improvement in the enumeration of urban-based indigenous people as well as the natality effect of rising indigenous male exogamy in cities (Taylor 2013). This urban growth aside, the more striking and profound observation concerning indigenous population distribution of the past 30 years (precisely because it has run counter to expectations and against standard theory of mobility transition) concerns the growth of remote indigenous towns (the former mission and government welfare settlements) alongside the increased dispersion of population on the indigenous estate. This has sustained a distribution of indigenous population that is very different to that overall as shown in Figure 6.2.

Reflecting their original widespread distribution, almost one-third of indigenous Australians remain in remote and very remote areas in contrast to a very small proportion of the rest of the population who tend to migrate in and out of such places in response to employment opportunities. In very remote areas (which account for almost 80 percent of the continent), indigenous people constitute almost half of the resident population, although on the indigenous estate away from scattered mining and service towns, they are very much the majority. In these areas an indigenous occupance mode has taken hold. Contrary to Rowley's (1971) idea of holding institutions preventing the urbanisation of indigenous people – if anything, it seems, they were being held

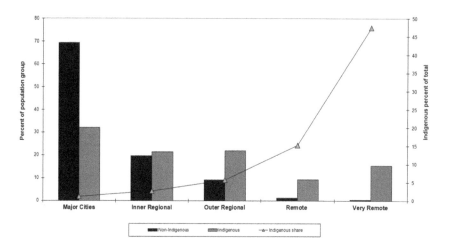

**Figure 6.2   Indigenous and Non-Indigenous Population Distribution by
                Remoteness Classification, 2006**

Source: ABS 2008

back from country since an immediate response to the dismantling of native welfare regimes in the early 1970s was the movement of family groups away from centralised polyglot settlements to reoccupy and reassert rights over traditional clan country. More recently, a demand for ecosystem services and the preservation of cultural landscapes have played a prominent role in this process (Holmes 2002, 2010, Altman and Kerins 2012). This desire to remain or move back to homelands, was well expressed by one Aboriginal women from Utopia, north of Alice Springs.

> All of the components of our identity hangs on the land. There's the land in a circle. There's the language from that land. It incorporates family lineage, family groups. It incorporates our sacred lands. It incorporates our law. The law is L.A.W. as well as L.O.R.E. Break any one of these arms and sever it from the land, you are committing the death of a race of people. (quoted in Marland 2012: 16)

While the initial shift to outstations was unsupported, the impetus gained by land rights led to the provision of Commonwealth-funded 'establishment' grants for boreholes and basic shelter to those who could demonstrate 'permanent' residence on country (a minimum of six months of the year) (Altman 2006: 5). So rapid and widespread was this migration that an official inquiry into outstations was established by a federal parliamentary standing committee in 1985 to report on circumstances and recommend a policy response (Commonwealth of Australia

1987). This revealed some 700 outstation communities on indigenous lands across the country with a population of almost 15,000. By 2006, the equivalent number of small (less than 1,000 persons) so-called 'discrete' indigenous settlements (locations with majority indigenous populations) was found to be 1,170 with a resident population of 65,650 (ABS 2007). Their distribution (Figure 6.3) reveals a strong association with the indigenous estate and marks both a rejection of the assimilation policies of earlier times and the outcome of Aboriginal initiatives, not those of the governments and officials (Marland 2012).

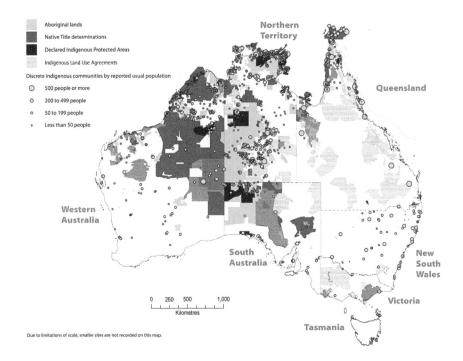

Figure 6.3   Discrete Indigenous Settlement and the Indigenous Estate

Source: Altman 2012: 11

There are clear concentrations of settlement on indigenous lands in the Northern Territory, across the Kimberley in Western Australia, in Cape York and the Torres Strait in Queensland and on the Anangu Pitjantjatjara Yankunytjatjara Lands in the north of South Australia and across the Western Australia border. Elsewhere, there are scattered indigenous communities on pockets of Aboriginal Lands Trust land across the south west of Western Australia and on lands held by Aboriginal Land Councils along the coast and across the interior of New South Wales. By contrast similar localities in much of interior Queensland are absent– these lands were mostly emptied of their original inhabitants by removals to larger government settlements towards the coast: Yarrabah, Palm Island, Woorabinda and Cherbourg (Anderson 1986).

The Northern Territory Experience

While the establishment of outstation settlement has clearly been widespread, the impact of this on population distribution has been difficult to track over time since such localities tend to be too small to be distinguished in the census (Taylor 1992). Indeed, it required a special survey of Indigenous Housing Organisations conducted by the ABS prior to each national census (the Community Housing and Infrastructure Needs Survey) to yield such information, this survey was disbanded after the 2006 census. However, in the Northern Territory where movement to outstations has been long standing and where administrative systems have been maintained by the Northern Territory government to monitor developments, it is possible to construct a time series covering the past two decades using local estimates of population at individual locations. This shows that while the number of very small family-based outstations has continued to proliferate, their average population size has declined to less than 10 persons (Table 6.1). The fact that the population in such places is now smaller than 20 years ago indicates that many of these have been effectively abandoned or only occupied intermittently which suggests that a lower limit to sustainable settlement size exists; indeed such a notion has always formed part of both Commonwealth and Territory policy approaches. Austin-Broos (2011:142) suggests that in the initial enthusiasm for decentralisation many people expected that town-based services would simply follow them and that even in the early 1990s in some central Australian outstations this included the idea that they might acquire their own shops and service stations. Not surprisingly, given market economics, this did not eventuate.

At the other extreme, the number and average size of indigenous towns (places over 1,000 persons) has substantially increased (Table 6.1). Twenty years ago there were only 3 such towns in the Northern Territory, by 2011 there were 11. Table 6.2 shows that the share of the indigenous estate population resident in these towns had risen over two decades from 12 percent to 34 percent with the actual number increasing by a staggering 362 percent (an annual rate of 7 percent). This represents a major shift in the nature of places that people live in and creates a rapidly changing set of opportunities for development and service delivery. It probably represents the highest rate of urban growth in rural and regional Australia, a fact that has not gone unnoticed by policy-makers.

Table 6.1   Number of Indigenous Settlements and Estimated Population by Settlement Size: Northern Territory Indigenous Estate, 1986 and 2008

Year	Settlement size							
	>1,000		500-999		50-499		0-49[1]	
	No.	Population	No.	Population	No.	Population	No.	Population
1986	3	3,809 (ave. size 1,270)	11	6,909 (ave. size 628)	112	14,784 (ave. size 132)	515	6,878 (ave. size 13)
2008	10	16,234 (ave. size 1,623)	9	6,345 (ave. size 705)	122	18,772 (ave. size 154)	607	5,738 (ave. size 9)

Note: 1. Includes locations with some infrastructure. In 1986, only 374 of the 515 such places listed had a recorded estimate of some resident population. In 2008, only 348 of the 607 listed had a recorded estimate of some resident population. As far as can be established all remaining localities in this category had no resident population.

Source: Taylor 2008

Table 6.2   Distribution of Population by Settlement Size on the
               Indigenous Estate: Northern Territory, 1986-2008

	>1,000	500-999	50-499	<50	Total
1986	11.8	21.3	45.7	21.2	32,380
2008	34.5	13.5	39.8	12.2	47,089
Percent change	326.2	-8.2	27.0	-16.6	45.4

Source: Taylor 2008

These data highlight two developments of note. First, there has been a maturation of the original 'return to country' sites at the base of the settlement hierarchy with population growth occurring *in situ* in the settlement size range 50-499 persons. By 2008, although movement to new locations continued, actual growth in population was confined to those places that were already established. Second, and most strikingly, there has been a process of incipient urbanisation. This is reinforced by a blurring in population composition between outstations and the larger townships that service them arising from the short-term movement of people into these towns on a daily or periodic basis to access essential services including health, schooling, and shopping, as well as for meetings, and social reasons. There is also strong inward seasonal movement during the wet season as roads to outstations can be impassable. Likewise, town residents frequently spend periods of time out bush during the dry season for recreational, ceremonial and social reasons. In effect, the townships and surrounding outstations on indigenous land tend to operate as population clusters in hub and spoke relationships such that estimates of service population, which can number up to 3,000 persons, are the more appropriate measure of settlement size especially in the more densely populated north of the Territory (Taylor 2004).

Emerging towns on the indigenous estate thus effectively play a similar servicing role to that of country towns elsewhere in Australia, except that they are poorly equipped to do so owing to substantial deficits in essential infrastructure. This is borne of their lack of an original economic base and subsequent neglect as part of the Australian settlement system (Dillon and Westbury 2007). The range of available services on the indigenous estate is related to settlement size but even urban places do not have the full range of services and infrastructure that would normally be found especially given an established trend of population growth. In 2006, there were notable shortfalls in hospitals, primary health care centres, preschools, secondary schools, aged accommodation, child care, youth centres, and (above all) housing (Table 6.3).

Table 6.3   Select Services and Infrastructure on the Indigenous Estate in the Northern Territory by Settlement Size Category, 2006

Population size	<100		100–499		500–999		>1,000		Total	
Pre-primary school	5	(0.9)	24	(38)	8	(67)	4	(67)	41	(6)
Primary school	56	(10)	46	(73)	12	(100)	6	(100)	120	(17)
Secondary school to Year 12	0	(0)	6	(9)	4	(33)	3	(50)	13	(2)
Aged accommodation	2	(0.3)	10	(16)	0	(0)	4	(67)	16	(2)
Store	11	(2)	47	(75)	12	(100)	6	(100)	76	(11)
Child care centre	1	(0.2)	24	(38)	9	(75)	5	(83)	39	(6)
Youth centre	0	(0)	15	(24)	4	(33)	5	(83)	24	(3)
Access to a doctor	106	(19)	44	(70)	11	(92)	6	(100)	167	(24)
Hospital	0	(0)	0	(0)	0	(0)	0	(0)	0	(0)
Primary health care centre	9	(2)	24	(38)	10	(83)	4	(67)	47	(7)
Percentage of dwellings requiring major repairs/replacement	27.8		30.2		40.3		29.5		31.2	
Occupancy rate per functional dwelling[1]	6.6		8.6		12.3		14.2		9.4	

Note: 1. Permanent dwellings not requiring major repair or replacement.

Figures in parenthesis indicate the percentage of communities in each size category with the selected service or infrastructure. Some of the communities listed here are town camps. This includes 32 of the 560 communities of less than 100 persons, and 10 of the 63 communities of between 100 and 499 persons.

Source: Taylor 2007

Almost one-third of dwellings in indigenous towns (all community rental at the time) required major repair or replacement leading to an average occupancy rate per dwelling of 14.2 persons.

Social Indicators

The task of generating social indicators for the indigenous estate awaits the matching of census and administrative data with land tenure boundaries. However, there are several regional and community studies that provide an indication of the key parameters, such as that for the Pilbara region in Western Australia where indigenous people have remained substantially marginalised despite being amidst the generation of vast wealth and economic activity associated with an expansionary mining sector (Langton 2012). Table 6.4 shows select socioeconomic outcomes for the region's adult indigenous population of 4,760 in 2006. The vast majority had limited schooling and post-school qualification; around half were not in the labour force; many were hospitalised at any one time; others were subject to chronic conditions requiring strict management regimes; many were arrested and incarcerated (especially young males); and feeding into this adult realm are relatively low achievers from the education system. In any event, the potential for prolonged and productive workforce participation is severely curtailed by premature mortality.

Table 6.4　　Summary Indigenous Social Indicators, Pilbara Region, 2006

Population aged 15+	4,759
Has no post-school qualification	4,200
Has less than Year 10 schooling	1,500
Not in the labour force	2,190
Hospitalised each year (all persons)	2,800
Has diabetes (25 years and over)	1,020
Has a disability	1,020
Arrested each year	1,050
In custody/supervision at any one time	310
Achieves Year 7 literacy (current attendees)	60%
15 year olds surviving to age 65	<50%

Source: Taylor and Scambary 2005

From a policy perspective, the levels of social and economic exclusion on the scale implied by such outcomes as being prevalent on the indigenous estate have inevitably raised questions about the adequacy of government resourcing and about the underlying structural causes of disadvantage and whether different approaches to servicing communities on the indigenous estate were necessary (Sutton 2009, Dillon and Westbury 2007). Morevoer such problems are exacerbated in larger Northern Territory towns, such as Wadeye, where indigenous groups from different areas are concentrated. By the turn of the century, the same questions began to be asked by governments.

Policy Responses

Recent analysis of competing principles in indigenous affairs policy suggests that a paradigm shift has been underway over the past few years involving a movement away from notions of choice and diversity and back towards earlier ideas of guardianship based on a conviction that governments can, and should (in order to achieve equity goals), intervene to shape and enhance Indigenous participation and life circumstances (Sanders 2008, 2009). Not surprisingly, this discourse has generated discussion regarding the viability of dispersed settlement on the indigenous estate. As a policy issue, this was first raised by a former federal indigenous affairs minister (Vanstone 2005) with more recent contributions arguing for the withdrawal of services entirely from locations that are not economically viable (Johns 2009). A critique is provided by Moran (2009). The actual government response has been complex, some would say conflicted (Altman 2012).

Since the late 1990s the signal from government to residents of the indigenous estate has been to eschew the possibilities of difference in order to embrace the institutions of mainstream Australian life with migration-inducing implications. Policy has emphasised the pursuit of market engagement with programs introduced to enhance labour mobility, increase home ownership, enact welfare to work reform, and encourage more individualised, as opposed to communal, articulation with government services. Combined with the evident lack of basic infrastructure provision and a growing population in towns that one former indigenous Northern Territory minister referred to as 'dysfunctional' (Ah Kit 2004), an accumulating mood for change led the Commonwealth government to exercise its constitutional powers in regard to the Northern Territory to enact a package of highly interventionist legislation in 2007 (referred to as the Northern Territory Emergency Response – NTER) designed to protect children and make communities safe. The immediate means to achieve this were various but they included measures to improve infrastructure and services and the acquisition of five year leases over townships to provide for commercial development and the introduction of market based rents and normal rental arrangements for what would become Territory-rental, instead of community-rental, housing. Significantly, a proportion of welfare payments were quarantined for certain goods and services for those resident in so-called 'prescribed communities' on indigenous lands and living areas (Altman and Hinkson 2007). The longer-term aims of the NTER (now taken up by the Commonwealth government's *Stronger Futures* policy) were focussed more on improving economic outcomes by raising the capacity of individuals to engage with mainstream labour market opportunities and improving their access to higher order services by concentrating public investment in a few strategic locations. All of these measures have potentially far reaching consequences for the emerging indigenous occupance mode.

Settlement Impacts of Welfare Reform

Two policy developments that emerged from the inquiry into the outstations movement in the 1980s assisted the expansion of dispersed settlement (Altman 2006: 6). The first was an extension of eligibility for the 'workfare' Community Development Employment Projects (CDEP) scheme to outstation residents. This effectively operated as an income support system enabling people to remain on country and access transfer payment income by overriding normal job search requirements whilst engaging in economic activities of their own choosing, many of which involved customary land and sea management activities and related pursuits, such as art production. The second was a more realistic assessment of infrastructure needs at outstations, especially access to adequate potable water and functional housing. Support for outstation residents inevitably spread to other service areas such as education, although only in a meagre fashion. In the context of more recent policy both of these props of remote settlement became threatened by the NTER and subsequent measures.

As part of the emergency response, CDEP scheme participants were to be transitioned off the scheme and placed on job search activities aimed at securing mainstream employment (Altman and Hinkson 2007). In this arrangement, the normal participation requirements of unemployed persons involving job search and formal training were to apply with eligible activity more narrowly defined than before, to exclude the more cultural and ad hoc activities that CDEP scheme participants had hitherto been able to engage in. From an indigenous community perspective, CDEP jobs had for a long time provided the scope for 'active' rather than 'passive' welfare via flexible arrangements for supporting a wide range of economic activities, mostly at the community's choosing. Under the new arrangements, an issue immediately arose about the composition of future government-funded work in the bush and how this was best determined – would it become just a route to mainstream work, or would it continue to operate as a means to rural occupance?

The universal provision of welfare cash payments, increasingly via the CDEP scheme, had over the past 40 years provided an adequate income for people to pursue indigenous agendas leaving them free to produce social and symbolic capital without the necessity for the great majority to be involved in material production in the wider economy, including the selling of labour. (Peterson 1999: 853) However, with the abandonment of CDEP and lifting of remote area exemptions on job search activity, a premium was placed on formal labour force participation and, given the spatial distribution of employment opportunities in favour of urban places, an inevitable stimulus for movement away from dispersed localities emerged. Added impetus for such movement is also now provided by plans to focus infrastructure development and service delivery in the Northern Territory's 20 largest indigenous towns whilst capping expenditure at outstations. This decision followed a 2007 review of the Community Housing and Infrastructure Program (CHIP) that argued that governments should desist from building new housing at outstations and instead provide 'mobility incentives' to assist families to move to larger settlements (PriceWaterhouseCoopers 2007). The former notion, at least, was taken up by the Northern Territory government's *Working Futures* policy – an avowedly centralising strategy aimed at building regional hubs, or 'growth towns'. While this has tended to be interpreted as an extension of the NTER (Kerins 2012), urban growth pre-dated the growth towns strategy to the extent that the new policy direction was as much a response to changing settlement dynamics as an attempt to influence them. Within this scheme, to qualify for any government funding an outstation has to be occupied for at least eight months of the year and funding levels were dependent on population size and distance to nearest hub. No new or additional funding was to be provided over and above current levels. Basically, the position adopted was that outstations are located on privately-owned lands under the *Aboriginal Land Rights (Northern Territory) Act 1976* and so, like other Australians who live on private land, indigenous residents should be responsible for their own repairs and maintenance of housing and infrastructure. Accordingly, statements of expectation of service delivery were to be drawn up for each outstation group. Given that such residents are among the poorest in the nation in terms of income and assets, the implied expectation, in many cases, must have been that they would leave.

However, in an apparent about-face, the Commonwealth government's *Stronger Futures* legislation introduced in 2011 pledged more than $221 million over ten years to be spent on outstations for essential services including access to power, water and sewerage as well as for municipal services such as road maintenance and garbage collection. Other spending was for extra ranger positions on country. This was tied to the declaration of IPAs as part of the National Conservation Estate and the establishment in 2007 of the Commonwealth-funded program *Working on Country* that now operates at some 90 sites nationally across the indigenous estate and employs over 680 indigenous rangers to undertake natural resource management work (Altman and Kerins 2012). These sorts of initiative raise questions about the strategic integrity of government policy in relation to remote living as it seems increasingly to pull in opposite directions. Further polarity in policy is evident in the recognition of growth towns and the quasi-local government servicing role that is envisaged for them under the *Working Futures* policy while at the same time community councils in these places have been dis-established in favour of a few overly-centralised 'super-Shires' (a reference to their vast incorporated areas). Interestingly, a change of government in the Northern Territory in 2012 reflected the voting patterns of constituencies on the indigenous estate who demanded a devolution of local government responsibilities and a return to localism (Sanders 2012). What, then, for the future of the indigenous occupance mode?

There is now some contradiction between renewed support for rural livelihoods, on the one hand, and policies that resonate with a set of national investment principles for the indigenous estate adopted as part of the *National Indigenous Reform Agenda* on the other. This latter declares that 'investment decisions in remote locations should aim to improve participation in education, training and the market economy (COAG 2009: A1). It also dictates that:

> ... priority for enhanced infrastructure support and service provision should be to larger and more economically sustainable communities ... recognising Indigenous peoples' cultural connections to homelands but avoiding expectations of major investment in service provision where there are few economic or educational opportunities ... and facilitating voluntary mobility by individuals and families to areas where better education and job opportunities exist with higher standards of services (COAG 2009: A1).

The conundrum for those working to these guidelines for service delivery and the achievement of social and economic equality goals is that support for Natural Resource Management may enhance formal employment outcomes but it also perpetuates locational disadvantage.

Conclusion

Indigenous values, aspirations, rights and development dilemmas form a growing component of a rural transition in Australia towards more multifunctional occupance. This is certainly true in terms of the large and growing share of land area that is subject to some form of indigenous proprietorship or stakeholder interest. This land transfer has helped to sustain an indigenous presence across rural and regional Australia in areas where population decline has otherwise been endemic. Associated with this has been a dispersal of population into small family-based settlements consistent with indigenous custodial relationships to land. Aspirations for this mode of occupance are likely to continue to spread, along with further expansion of the indigenous estate, as long as government support for essential infrastructure and services occurs alongside any independent finance for such that might derive from private agreement making, for example with resource developers. At the same time, populations in hub towns have also expanded and this trend seems set to continue given a policy preference for centralised services.

According to Holmes (2010: 351), external pressures are likely to ensure further functional change in these areas with indigenous occupance modes and trajectories becoming more closely aligned to non-indigenous ones in comparable regional contexts. This reflects the vulnerability of the indigenous occupance mode owing to its dependence on government transfers and the program imperatives of federal, state and territory policies. The idea that such workings of political economy might produce vulnerability for particular populations and locations is emphasised in spatial analyses that attempt to 'place' vulnerability by investigating the spatial contexts and ways in which social relations produce 'vulnerable spatialities' (Findlay 2005, Philo 2005). Indigenous populations and the spaces they occupy are (almost by definition) vulnerable owing to their special relations of dependency to a regional or national core.

Mainstream analysis and discourse on rural and regional development has tended to overlook this indigenous presence and, as a consequence, significant population dynamics and development priorities are rendered invisible (Pritchard 2003). This hinders an integrated analysis of rural transition just at a time when there is questioning of the degree to which governments are prepared to support indigenous peoples who choose to reside in dispersed settlements on lands that they own. Put simply, there is an assessment by the state that life chances will be improved by a resettlement of indigenous people into fewer larger locations that have more services and mainstream opportunities. This view effectively returns policy to its position of 40 years ago and it has stimulated a debate as to whether the goal of indigenous policy should be to achieve equality of socioeconomic status or to facilitate choice and self-determination (Austin-Broos 2011, Altman and Rowse 2005) – the former tends to imply integration and urban-ward migration, while the latter tends towards more multifunctional occupance. A spatial conundrum has therefore developed in indigenous regional development – land rights bestow a legitimate interest in residence away from the mainstream economy and institutions, but access to these is considered by government to be the key means to 'closing the gap'.

References

Ah Kit, J., 2004, 'Why Do We Always Plan for the Past? Engaging Aboriginal People in Regional Development', Paper to the *Sustainable Economic Growth for Regional Australia (SEGRA) 8th National Conference*, Alice Springs, 7th September.

Altman, J.C., 2001, 'Sustainable Development Options on Aboriginal Land: The Hybrid Economy in the 21st Century', *CAEPR Discussion Paper No. 226*. Canberra: Centre for Aboriginal Economic Policy Research, ANU.

Altman, J.C., 2006, 'In Search of an Outstations Policy for Indigenous Australians', *CAEPR Working Paper No. 34*, Canberra: Centre for Aboriginal Economic Policy Research, ANU.

Altman, J.C., 2012, 'Land Rights and Development in Australia: Caring For, Benefiting From and Governing the Indigenous Estate', in L. Ford and T. Rowse (eds), *Between Indigenous and Settler Governance*, London: Routledge, 121-134.

Altman, J.C., Buchanan, G.J. and Larsen, L., 2007, 'The Environmental Significance of the Indigenous Estate: Natural Resource Management as Economic Development in Remote Australia', *CAEPR Discussion Paper No. 286*, Canberra: Centre for Aboriginal Economic Policy Research, ANU.

Altman, J.C. and Kerins, S. (eds), 2012, *People on Country: Vital Landscapes/ Indigenous Futures*, Sydney: The Federation Press.

Altman, J.C. and Pollack, D., 2001, 'The Indigenous Land Corporation: An Analysis of its Performance Five Years On', *Australian Journal of Public Administration*, 60: 67-79.

Altman, J., and Hinkson, M. (eds), 2007, *Coercive Reconciliation: Stabilise, Normalise, Exit Aboriginal Australia*, Melbourne: Arena Publications.

Altman, J. and Rowse, T., 2005, 'Indigenous Affairs', in P. Saunders and J. Walter (eds), *Ideas and Influence: Social Science and Public Policy in Australia*, Sydney: UNSW Press, 159-177.

Anderson, C., 1986, 'Queensland Aboriginal Peoples Today', in J.H. Holmes (ed.), *Queensland: A Geographical Interpretation*, Brisbane: Queensland Geographical Journal, 296-320.

Austin-Broos, D., 2011, *A Different Inequality: The Politics of Debate about Remote Aboriginal Australia*, Sydney: Allen and Unwin.

Australian Bureau of Statistics (ABS), 2007, *Housing and Infrastructure in Aboriginal and Torres Strait Islander Communities, Australia, 2006*, Canberra: ABS.

Australian Bureau of Statistics (ABS), 2008, *Experimental Estimates of Aboriginal and Torres Strait Islander Australians, June 2006*, Canberra: ABS.

Australian Human Rights Commission (AHRC), 2011, *Native Title Report 2011*, Sydney: Aboriginal and Torres Strait Islander Social Justice Commissioner, AHRC.

Baker, R., Davies, J. and Young, E., 2001, *Working on Country: Contemporary Indigenous Management of Australia's Lands and Coastal Regions*, Melbourne: Oxford University Press.

Biddle, N., 2009, 'The Geography and Demography of Indigenous Migration: Insights For Policy And Planning', *CAEPR Working Paper No. 58*, Canberra: Centre for Aboriginal Economic Policy Research, ANU.

Commonwealth of Australia, 1987, *Return to Country: The Aboriginal Homelands Movement in Australia*, Canberra: Australian Government Publishing Service.

Council of Australian Governments (COAG), 2009, *National Partnership Agreement on Remote Service Delivery*, Canberra: COAG.

Crough, G. and Christophersen, C., 1993, *Aboriginal People in the Economy of the Kimberley Region*, Darwin: North Australia Research Unit, ANU.

Davies, J., 2003, 'Contemporary Geographies of Indigenous Rights and Interests in Rural Australia', *Australian Geographer*, 34: 19-45.

Dillon, M.C. and Westbury, N., 2007, *Beyond Humbug: Transforming Government Engagement with Indigenous Australians*, Westlakes, South Australia: Seaview Press.

Findlay, A., 2005, 'Editorial: Vulnerable Spatialities', *Population Space and Place*, 11: 429-39.

Gerritsen, R., Stanley, O., and Stoeckl, N., 2010, 'The Economic Core? The Aboriginal Contribution to the Alice Springs/Central Australian Economy', *Journal of Economic and Social Policy*, 13(2). Available at http://epubs.scu.edu.au/jesp/vol13/iss2/5 [accessed:23/10/2012].

Holmes, J., 2002, 'Diversity and Change in Australia's Rangelands: A Post-productivist Transition with a Difference?', *Transactions of the Institute of British Geographers*, 27: 362-84.

Holmes, J., 2010, 'Divergent Regional Trajectories in Australia's Tropical Savannas: Indicators of a Multifunctional Rural Transition', *Geographical Research*, 48: 342-358.

Johns, G., 2009, 'No Job No House: An Economically Strategic Approach to Remote Indigenous Housing', Canberra: The Menzies Research Centre.

Kerins, S., 2012, 'Caring for Country to Working on Country', in J. Altman and S. Kerins (eds), *People on Country: Vital Landscapes/Indigenous Futures*, Sydney: The Federation Press, 26-44.

Langton, M., 2012, 'The Resource Curse Compared: Australian Aboriginal Participation in the Resource Extraction Industry and Distribution of Impacts', in M. Langton and J. Longbottom (eds), *Community Futures, Legal Architecture: Foundations for Indigenous Peoples in the Global Mining Boom*, Oxford and New York: Routledge, 23-44.

Langton, M., Rhea, Z.M., and Palmer, L., 2005, 'Community Oriented Protected Areas for Indigenous Peoples and Local Communities', *Journal of Political Ecology*, 12: 23-50.

McKenzie, F., 1994, *Regional Population Decline in Australia*, Canberra: Australian Government Publishing Service.

McGuirk, P. and Argent, N., 2011, 'Population Growth and Change: Implications for Australia's Cities and Regions', *Geographical Research*, 49: 317-335.

Marland, S., 2012, 'The Need for Homelands', *Australian Quarterly*, 83(3): 16-20.

Moran, M., 2009, 'What Job, Which House? Simple Solutions to Complex Problems in Indigenous Affairs', *Australian Review of Public Affairs* [Online]. Available at http://www.australianreview.net/digest/2009/03/moran.html [accessed: 24/10/2012].

National Institute for Rural and Regional Australia (NIRRA), 2012, 'Scoping a Vision for the Future of Rural and Regional Australia'. Canberra: NIRRA, ANU. Available at http://www.nirra.anu.edu.au [accessed: 25/10/2012].

O'Faircheallaigh, C., 2012, 'Curse or Opportunity? Mineral Revenues, Rent Seeking and Development in Aboriginal Australia', in M. Langton and J. Longbottom (eds), *Community Futures, Legal Architecture: Foundations for Indigenous Peoples in the Global Mining Boom*, Oxford: Routledge, 45-59.

Pearson, N., 2000, *Our Right to Take Responsibility*, Cairns: Noel Pearson and Associates.

Peterson, N., 1999, 'Hunter-gatherers in First World Nation States: Bringing Anthropology Home', *Bulletin of the National Museum of Ethnology*, 23: 847-61.

Philo, C., 2005, 'The Geographies that Wound', *Population Space and Place*, 11: 441-54.

Pollack, D., 2001, 'Indigenous Land in Australia: A Quantitative Assessment of Indigenous Landholdings in 2000', *CAEPR Discussion Paper No. 221*, Canberra: Centre for Aboriginal Economic Policy Research, ANU.

PriceWaterhouseCoopers, 2007, *Living in the Sunburnt Country. Indigenous Housing: Findings of the Review of the Community Housing and Infrastructure Needs Programme*, Canberra: Department of Families, Community Services and Indigenous Affairs.

Pritchard, B., 2003, 'Indigenous People and Australia's Regions', in A. Beer, A. Maude and B. Pritchard (eds), *Developing Australia's Regions: Theory and Practice*, Sydney: UNSW Press, 171-191.

Rowley, C.D., 1971, *The Remote Aborigines*, Canberra: Australian National University Press.

Rowse, T., 2012, *Rethinking Social Justice: From 'Peoples' to 'Populations'*, Canberra: Aboriginal Studies Press.

Sanders, W., 2008, 'In the Name of Failure: A Generational Revolution in Indigenous Affairs', in C. Aulich and R. Wettenhall (eds), *The Fourth Howard Government*, Sydney: UNSW Press, 187-205.

Sanders, W., 2009, 'Ideology, Evidence and Competing Principles in Australian Indigenous Affairs: From Brough to Rudd via Pearson and the NTER', *CAEPR Discussion Paper No. 289*, Canberra: Centre for Aboriginal Economic Policy Research, ANU.

Sanders, W., 2012, 'Winning Aboriginal Votes: Reflections on the 2012 Northern Territory Election', *Australian Journal of Political Science*, 47: 691-701.

Steering Committee for the Review of Government Service Provision (SCRGSP), 2011, *Overcoming Indigenous Disadvantage: Key Indicators 2011*, Canberra: Productivity Commission.

Sutton, P., 2009, *The Politics of Suffering: Indigenous Australia and the End of the Liberal Consensus*, Melbourne: Melbourne University Press.

Taylor, J., 1992, 'Geographic Location and Economic Status: A Census-based Analysis of Outstations in the Northern Territory', *Australian Geographical Studies*, 30: 163-84.

Taylor, J., 2003, 'Indigenous Australians: The First Transformation', in S. Khoo and P. McDonald (eds), *The Transformation of Australia's Population: 1970-2030*, Sydney: UNSW Press, 17-39.

Taylor, J., 2004, *Social Indicators for Aboriginal Governance: Insights from the Thamarrurr Region, Northern Territory*, CAEPR Research Monograph No. 24, Canberra: ANU E Press.

Taylor, J., 2007, 'Demography is Destiny – Except in the Northern Territory', in J. Altman and M. Hinkson (eds), *Coercive Reconciliation: Stabilise, Normalise, Exit Aboriginal Australia*, Melbourne: Arena Publications, 173-185.

Taylor, J., 2008, 'The Demography of NTER Prescribed Areas', in P. Yu, M. Duncan and B. Gray (eds), *Northern Territory Emergency Response Report of the NTER Review Board*, Canberra: Commonwealth of Australia, 92-98.

Taylor, J., 2013, 'Indigenous Urbanization in Australia: Patterns and Processes of Ethnogenesis', in E. Peters and C. Andersen (eds), *Indigenous in the City: Contemporary Identities and Cultural Innovation*, Vancouver: University of British Columbia Press, 237-255.

Taylor, J. and Bell, M., 2004, 'Continuity and Change in Indigenous Australian Population Mobility', in J. Taylor and M. Bell (eds), *Population Mobility and Indigenous Peoples in Australasia and North America*, London: Routledge, 13-43.

Taylor, J. and Biddle, N., 2010, 'Estimating the Accuracy of Geographic Variations in Indigenous Population Counts', *Australian Geographer*, 41: 469-485.

Taylor, J. and Scambary, B., 2005, *Indigenous People and the Pilbara Mining Boom: A Baseline for Regional Participation*, CAEPR Research Monograph No. 25, Canberra: ANU E Press.

Taylor, J. and Stanley, O., 2005, 'Opportunity Costs of the Status Quo in the Thamarrurr Region, Northern Territory', *CAEPR Working Paper No. 28*, Canberra: Centre for Aboriginal Economic Policy Research, ANU.

Tehan, M. and Godden, L., 2012, 'Legal Forms and their Implications for Long-term Relationships and Economic, Cultural and Social Empowerment: Structuring Agreements in Australia', in M. Langton and J. Longbottom (eds), *Community Futures, Legal Architecture: Foundations for Indigenous Peoples in the Global Mining Boom*, Oxford: Routledge, 111-133.

Trigger, D., 1986, 'Blackfellas and Whitefellas: The Concepts of Domain and Social Closure in the Analysis of Race Relations', *Mankind*, 16: 99-117.

Vanstone, A., 2005, 'Beyond Conspicuous Compassion', *Address to the Australia and New Zealand School of Government*, ANU, Canberra, December.

Von Sturmer, J.R., 1984, 'The Different Domains', in Australian Institute of Aboriginal Studies, *Aborigines and Uranium: Consolidated Report to the Minister for Aboriginal Affairs on the Social Impact of Uranium Mining on the Aborigines of the Northern Territory*, Canberra: Australian Government Publishing Service.

Chapter 7

Challenging Mining Workforce Practices: Implications for Frontline Rural Communities

Alison McIntosh and Kerry Carrington

Introduction

Global demand for minerals and energy products has fuelled Australia's recent 'resources boom' and led to the rapid expansion of mining projects not solely in remote regions but increasingly in long-settled traditionally agriculture-dependent rural areas. Not only has this activity radically changed the economic geography of the nation but a fundamental shift has also occurred to accommodate the acceleration in industry labour demands. In particular, the rush to mine has seen the entrenchment of workforce arrangements largely dependent on fly-in, fly-out (FIFO) and drive-in, drive-out (DIDO) workers. This form of employment has been highly contentious in rural communities at the frontline of resource sector activities.

In the context of sweeping structural changes, the selection of study locations was informed by a range of indices of violence. Serendipitously we carried out fieldwork in 2008 in communities undergoing rapid change as a result of expanding resource sector activities. The presence of large numbers of non-resident FIFO and DIDO workers was transforming these frontline communities. This chapter highlights some implications of these changes, drawing upon one particular location, which historically depended on agriculture but has undergone redefinition through mining.

A Snapshot of Australia's 'Resources Boom'

The massive ramp-up in resource sector activities responsible for Australia's resources boom can be illustrated through expenditure on the construction of new and expansion projects to increase production and exports. In the six months to April 2012, completed new projects represented 22 percent of the total worth of all projects for the six years to that date ($107.5 billion in 2011-12 dollars). Moreover, the value of 'advanced' projects – those underway but not completed – was $261 billion, a staggering 6.5 times greater than six years previously (BREE 2012). Additionally, projects moving through the planning and approvals pipeline have remained valued at about $250 billion since 2009. Irrespective of the continuance or decline in new project development, the value of production is expected to increase further, repeating historic patterns in Australian mining industry GDP between bursts of development activity (ABS 2012a).

This recent rush to mine has created huge labour demands. During the 1960s and 1970s, an earlier period of intense mining activity, a condition of development consent for new projects in regional and remote areas was for mining companies to purpose-build towns for workers and their families (Houghton 1993). Since the late 1970s, this 'new town' approach has largely been replaced by a 'no town' model (Storey 2010) although until recently, in all but the remotest locations, operational workers were mainly local residents.

Given the accelerated demands for labour, the reliance on FIFO and, more recently, DIDO and even BIBO (bus-in, bus-out) and SISO (ship-in, ship-out) workers has intensified. The practice of using these types of non-resident workers (NRWs) itself is not new: in the Australian mining industry, FIFO has been around since the early 1980s, but the switch to these practices as the norm has besieged frontline communities. Whereas in 1970 all mining industry workers in Western Australia lived locally, by 2005 only 53 percent did so (WACME 2005) and nowadays virtually all new recruits are NRWs. Furthermore, mining companies are distanced from many workforce issues through their engagement of principal contracting companies for project works. For instance, more than half (54 percent) of those engaged in the day-to-day business of mining in Central Queensland's Bowen Basin in mid-2011 were outside contractors (OESR 2012). Geographic remoteness and limited mine life are no longer principal drivers – as they once were – in corporate decisions to depend largely on NRWs. Moreover proponents of advanced and new mining projects have signalled that they will be largely reliant on NRWs in the future (Deloitte Access Economics 2011, NRSET 2010, URS 2012).

In spite of heightened resource sector activity, workers in 'mining' represent only a small proportion of the nation's workforce – 3.3 percent in May 2012 – although the number has more than doubled (to 266,600 persons) over the last five years (ABS 2012b). However, only extraction work from operating mines, some support activities and exploration are counted under the ANZSIC mining industry classification. Countless tens of thousands of construction workers and unidentified numbers for surveying, transportation, processing, out-sourced plant maintenance and work camp operations are not included. Because these types of workers are largely contracted NRWs, the industry's workforce and FIFO/DIDO growth rate is substantially greater than for 'mining' alone. Forecasts conceal true workforce size by counting only those NRWs in the work cycle of rosters and leaving out those away on leave. This means that, for even-time rosters with the same number of days/weeks in the work and leave cycles (for example, two weeks on, two weeks off), the total number of transient NRWs is double the number nominated. Accordingly, the true size of the sector's workforce remains speculative in the absence of industry disclosure. A conservative estimate for July 2011 put the number of NRWs at 150,000-200,000 NRWs (Carrington *et al.* 2011), but, relatively little is known about NRWs and their particular significance.

Workplace Arrangements for FIFOs/DIDOs

For all types of jobs within the industry, 12-hour shifts rotating between night and day have become the norm. Rosters vary greatly although two weeks on followed by one week off is common for mining operations; for construction, one week's leave typically follows a four-week work cycle. Industry incomes are well above average national earnings (ABS 2008) since high pay rates are essential to attract workers to regional and remote locations – or to poach from other industries, government or competitors – and to compensate for difficult, sometimes extreme, working and living conditions.

During work cycles, NRWs usually stay in work camps which have varying levels of amenity and can cater for up to several thousand at a time. Whereas once these camps were found only in remote locations, increasingly they are near long-established towns.

NRWs and work camps often become highly contentious especially where the local population is augmented by tens of thousands, mostly men. For instance, in Queensland's Bowen Basin, NRWs in the population grew by 40 percent in the 12 months to July 2011 to 20,520 extending the local resident population by 24 percent (OESR 2012). Nor are NRWs evenly dispersed within affected regions and thus residents in locations with very high concentrations feel the effects of these workforce practices more intensely. Examples of frontline communities experiencing conflicts include Port Hedland, Karratha and Geraldton in Western Australia's Pilbara and Midwest Regions; Coppabella, Moranbah, Middlemount, Moura, Miles and Chinchilla in Queensland's Bowen and Surat Basins; and Muswellbrook, Boggabri and Narrabri in the New South Wales Hunter Valley and Gunnedah Basin. Most of the recent intense development activity has occurred in, and is planned for, Western Australia and Queensland. Within these jurisdictions, the exceptionally dynamic regions, when measured by existing and planned operations, are the Pilbara Region in the west and Central Queensland's Bowen Basin. Our case study is within the latter region.

Burdens Unevenly Borne by Frontline Communities

While the high incomes paid to resource sector workers are enjoyed by relatively few, the social impacts of mining are also limited and unevenly dispersed. Costs are mostly borne by frontline communities which can become overwhelmed by the rapid influx of large numbers of NRWs and work camps. Moreover, NRWs generate significant gender imbalance because men represent around 85-90 percent of this worker cohort (ABS 2012b). These work practices and characteristics – the transient predominantly male workforce, dense work schedules, long shifts, and proliferation of work camps of variable standards – carry significant risks for rural communities. Rising levels of alcohol-fuelled violence, higher rates of drug abuse, gambling and other forms of conspicuous consumption and indebtedness, rising rates of fatigue-related death and injuries, erosion of health and wellbeing, soaring housing costs and other costs of living, and stretched basic services and infrastructure are among the downsides of these industry-entrenched workplace arrangements (Pini, McDonald and Mayes 2012, Carrington, Hogg and McIntosh 2011, Carrington, McIntosh and Scott 2010, Langton 2010, Murray and Peetz 2010, Haslam McKenzie *et al.* 2008). Identified impacts diminish social, human, institutional and environmental capital and undermine community stability and solidarity and the sustainability of frontline towns (Cleary 2012, 2011, Lozeva and Martinova 2008, Mayes 2008, Watts 2004).

Costs to communities have largely escaped industry, government and academic scrutiny, although the Australian Government commenced an inquiry into the effects of FIFO/DIDO workforce practices in regional Australia in 2011. Even anticipated economic benefits are not generally experienced within frontline communities due to 'fly-over' effects, with local businesses bypassed (Haslam McKenzie 2010, Storey 2010, 2001) and jobs shifting to NRWs to the extent of locals being excluded altogether by some mining corporations. The new BHP Mitsubishi Alliance Caval Ridge mine in the Bowen Basin, for example, is to have a 100 percent FIFO workforce. These arrangements are also influenced by Federal Government taxation policies which provide substantial economic benefits to companies favouring NRWs accommodated in camps. For example, companies can write off capital expenditure on camps as development costs and be exempt from fringe benefits tax for NRW transport and living expenses whereas subsidised housing and other living expenses for resident workers are taxed.

We considered over 40 Local Government Areas (LGAs) within regional and remote New South Wales, Queensland and Western Australia as potential study locations prior to identifying three for intensive field work, one in each of these states. As part of this selection process, national databases were analysed and then sifted for violence-related crime, mortality, morbidity, injury and accident data (such as violent crimes, motor vehicle fatalities, firearms injuries, and other injury surveillance data) and linked with socio-demographic characteristics by gender for persons living in regional and remote Australia. The LGAs chosen for field work contained a number of towns and villages and all had resident populations at the time of the 2006 Census of around 15,000 persons.

Primary qualitative data were gathered through semi-structured mostly one-on-one interviews with 142 civic and opinion leaders, community representatives, justice, human and medical service providers, industry representatives and workers. We explored perspectives on rural men's health and wellbeing, experiences of violence-related harm as victims and perpetrators, and related policy issues. In the Bowen Basin (Central Queensland) LGA, a strategic agricultural area and resource rich region which has, over several decades, been transformed through mining activities, 48 respondents were interviewed and selected results are presented here. The sensitivity of the research topic influenced our use of typical case sampling and the targeting of persons who were regarded as representative of the wider population. To preserve the anonymity and confidentiality of informants in low-population rural locations where community or professional roles can often identify individuals, pseudonyms are used for the LGAs and locations within study areas. The following draws upon our extensive qualitative research in the 'Bramely' LGA of the Bowen Basin.

'Bramely' Case Study

Agriculture remains important in 'Bramely'. Most mining operations and related activities are within the 'Serveton' area (with 49 percent of LGA population) which includes 'Canopus', the long-established main service town and population centre for 'Bramely', and 'Denebola', a nearby 'dormitory' town for resident and DIDOs/FIFOs employed in the resources sector. The geographically larger 'Baredge' area was home to the remaining LGA population with residents living and working on rural holdings as well as in small towns and villages.

Selected 2006 Census statistics for the 'Bramely' LGA are presented in Tables 7.1 and 7.2. Numbers of NRWs who were in the LGA on census night cannot be identified although most of the recorded visitors were probably DIDOs/FIFOs. Both Serveton and Baredge had comparatively high proportions of visitors (Table 7.1).

If characteristics for NRWs were included for 'Bramely', then the gender imbalance towards males would magnify the 'blokeyness' of the area and show a trend towards more single, middle-aged males and fewer family structures, including a scarcity of grandfathers. When asked if trouble arose in town between non-resident men and locals due to gender imbalance, one law enforcement officer in Serveton commented:

> For sure there's a definite disparity there with the amount of males and females in town. I wouldn't know exactly but I'd say there'd at least be four or five to one in town, to the side of the men. Maybe there's a little bit of a problem with that but there's probably a lot of town people probably

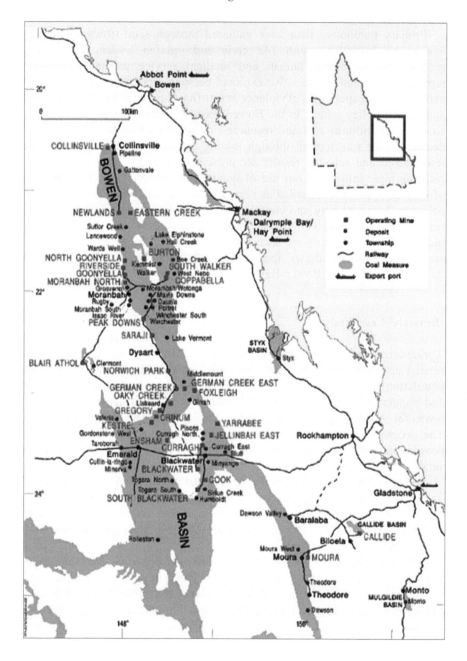

Figure 7.1   Central Queensland's Bowen Basin Coal Measures

Source: MISHC (2012, online)

Table 7.1   Visitor Population for 'Bramely' LGA

Characteristics	Serveton	Baredge	Australia
Persons visiting on Census night:	12.1	12.2	4.7
15-24 years	11.6	18.2	5.4
25-34 years	13.8	16.5	5.3
35-44 years	16.0	13.4	4.1
45-54 years	16.9	15.0	4.4
55-64 years	21.0	15.6	6.1

Source: ABS 2006 Census of Population and Housing

> wouldn't want their partners going to the hotel unaccompanied; the fear, there's lots of single blokes there looking for women. Also I've spoken to a lot of people, particularly people that are involved with the Pony Club – the Pony Club's just on the other side of the single men's camp – they won't let their daughters walk out to feed and water their horses because they're spot under [the eyes of] the single men in town ...

In Serveton, the 12-hour shifts worked by resident miners impacted upon levels of volunteering which, although substantially lower than for agricultural Baredge, were still greater than for Australia overall. A stand-out feature for Serveton was the comparatively high level of population mobility. In contrast, mobility of males in Baredge was more aligned with the Australian norm.

One in three Serveton males was employed in mining (Table 7.2) while agriculture was the most important industry in Baredge. The high proportion of machinery operators and drivers, technicians and trade workers in 'Bramely' aligns with employment in the resources sector. Another notable feature was the high weekly incomes, in excess of $2,000, earned by many males. At the same time, few young adult males in 'Bramely' were furthering their education, perhaps assuming they would get a mine job that did not require tertiary qualifications.

Evidence from other contexts indicates that the Bowen Basin is similar to many other frontline communities affected by resource sector activities. The situation in 'Bramely', therefore, has widespread relevance for industry, governments and communities in Australia and globally.

As far as our local informants were concerned, resource sector non-resident workforces were eroding community wellbeing and diminishing social capital. NRWs can have a destabilising impact upon frontline mining communities because:

Table 7.2   Selected Characteristics of Adult Males in 'Bramely' (percent)

Characteristics for adult males	Serveton	Baredge	Australia
Adult males employed in:			
Agriculture, fishing & forestry	1.9	42.7	4.0
Mining	32.2	13.1	1.8
Manufacturing	12.6	6.1	14.4
Utilities	7.9	2.6	1.4
Construction	9.0	8.0	12.5
Wholesale trade	2.8	3.1	5.3
Retail trade	7.1	2.9	9.0
Accommodation & food services	2.7	1.9	5.0
Public admin & safety	4.2	5.5	7.0
Education & training	1.8	1.5	4.3
Health care/ social assistance	1.4	1.0	4.2
Other	13.7	9.3	28.3
Occupations of males:			
Managers	8.3	38.3	16.1
Professionals	8.0	3.1	17.3
Technicians/trades workers	30.6	15.1	22.7
Machinery operators/drivers	26.3	20.1	11.0
Labourers	17.3	16.1	12.3
Other	9.5	7.3	20.6
Males attending an educational institution:			
15-19 years	43.5	43.8	66.0
20-24 years	6.8	4.1	29.5
Males with Incomes of:			
less than $250 per week	14.1	19.4	24.6
$1,000 or more per week	43.2	27.1	25.1
$2,000 or more per week	12.8	7.8	5.6

Note: 'Adult' refers to a male aged 15 years or over.

Source: ABS 2006 Census of Population and Housing

They never belong to the town; they never have responsibilities in the town; they never become involved or committed to the town ... Their families are elsewhere; it never becomes their town and I don't think that's good for the families.

A resident employee of a mining company recounted some of the social impacts of increasing numbers of NRWs in the community:

They [the NRWs] just come in, do their job ... You see that when there's vandalism and things around, when there's fights at the pub. Big shutdowns on, they finish, they have a few drinks. Yeah, we've seen it a few times here; we've had to change our schedules and work out better ways how to do shuts and your operations too and stop that [violence]. Well basically if they're working long hours they get fatigued and then they drink and the mixture is an explosion happening. They're frustrated, they want to get home to their partners so that can be taken out, you know. There's even things in town now, like, they have barmaids that are topless, you know, things like that being integrated back into town. When I first moved here, they didn't have that and I find it quite strange because that creates the culture of 'Sunday night, it's girlie night at the pub'.

Locals expressed a sense of grievance that NRWs disrespected and lacked commitment to the local community as this resident's comment illustrates:

That's the problem ... They [the NRWs] just don't care ... They're highly paid, um, the majority of them are not highly skilled; like, there're guys you wouldn't employ in town who get jobs out there [at the mine] on 100,000 [dollars] a year, driving trucks that are huge, you know, potential massive injuries if something happens.

Another participant explained how he thought social divisions were created by mining 'boom' conditions:

You've got the whole mining thing. You've got the real estate, it goes up, the rental goes up. The miners can afford that – or the mining companies who are paying the rentals on behalf of their staff, you know – and that creates resentment in the town. If people have just got ordinary trade jobs in the town or fifty thousand a year jobs, all of a sudden they've got to be paying $300 a week rent which is a bit of a struggle, if they can find it in the first instance.

Most people readily recalled recent violence involving NRWs, principally male-on-male assaults. Many considered this was the norm, given the dynamics between 'locals' and NRWs who were colloquially referred to as 'contractors'. As one local government councillor remarked: 'There've been a few bashings of recent times ... You're going to get a bit of violence with a bit of alcohol and contractors, you know, that's always on the cards'. Not only were NRWs largely blamed for violence involving locals but violence between some NRWs also commonly occurred. The manager of one work camp recalled:

> One of the most recent [fights] that I know of was where people that were within their own company, like contractors here for only a short period of time … It's like: We're all here to work and there's one slacker sponging off everyone else; or might have had a bad day at work and they go to the pub afterwards …

Pub brawls were fairly common in Denebola, the dormitory town for the resource sector. One respondent, a senior nursing practitioner who regularly attended to injuries such as 'glassings', where broken bottles and glasses are used as weapons, thought that NRWs were 'a little bit more aggressive than the local people':

> They don't have any rhyme or reason for having an argument; it just seems that they get a bit of alcohol in them and they have a fight … They come up here [to the hospital] and we patch them up and then we probably don't see them again. They go to the private practice after that or they go back to their town that they came from.

Levels of antagonism between locals and the NRWs escalated when the number of NRWs in the community rapidly increased. One police officer commented:

> Yeah, there's still a fair bit of violence around the hotels, especially with the contractor versus town person mentality. That's always been around where you've got the people who claim to be from Denebola and they're probably against contractors being in town and there's a predominance of contractors in town.

Although NRWs were usually cast as the initiators of violence, the workplace/lifestyle regimes imposed upon them attracted sympathy from some locals, including this local politician: 'They're in the camps, you know … They've got to stay in these little boxes, you know? These little dongas …'. Many we interviewed recognised that not all NRWs were violent, that some offenders were worse than others, and there were ways of diffusing confronting situations. One prominent businessman commented:

> A lot of the community safety issues around violence at the moment get pinpointed to contractors … There was a crew of contractors that was here that was fairly violent. They stayed at the pubs fairly regularly and the police had their contracting company actually split them up and sent some elsewhere. Otherwise they were all going to end up in jail so there were some fairly heavy violent incidents.

Focussing attention on the poor behaviour of some NRWs deflected criticism about equivalent conduct among the locals. Some local people did point out that the outside contractor was not always the initial instigator of violent behaviour. A senior police officer mused:

> ... [it's] quite funny really because half the locals try to get a job as a contractor as well yet they'll still go to the hotel and have a go at a contractor ... I s'pose the major definition is people that live here; and people that don't live here.

Thus the metaphorical battle line was not between contractors employed in the resources sector and residents but rather between NRW 'outsiders' and locals, the 'insiders'. In Western Australia too, the locals felt invaded and threatened by the men living in camps in close proximity to their town (Carrington *et al.* 2012, Carrington, Hogg and McIntosh 2011, Scott, Carrington and McIntosh 2011). A common strategy there was to construct outsiders as responsible for local crime problems, deflecting attention from the social dynamics of the existing or wider culture of violence in rural Australia. Irrespective of the reality, 'the violence and social disorder associated with drunken men from the work camps created a climate of fear and anxiety about safety' (Carrington, McIntosh and Scott. 2010: 405). This social dynamic was replicated in Serveton, as one long-term businessman remarked: 'If an argument starts, violence starts, community safety decreases. A lot of the community safety issues around violence at the moment gets pinpointed to contractors'.

Others we interviewed played down concerns about levels of violence. Interestingly, these opinions were generally qualified by making comparisons with social norms during Serveton's earlier mining years or, alternatively, with other mining towns within the region. For example, a mine union official and long-term resident thought that levels of violence were now: 'Nothing out of the ordinary. Like, when we have shutdowns, you might have a lot of contractors in there and you have a little bit of animosity but there's only little bits here and there'.

The reason for hostilities was sometimes presented as conflict between those associated with 'old' rural and 'new' mining. Mining and energy operations had been active in the Serveton area for several decades. Accordingly, levels of violence in Canopus and Denebola were less than in other towns within the region where more recent rushes to develop new mines have been accompanied by the huge influx of transient workers. According to one hotel manager, antagonism between established rural residents and the new cohort of FIFOs/DIDOs was responsible for:

> ... definitely a lot more hotel-night club violence ... The 'us and them' thing tended to be [between] the miners and everybody else regardless of whether the miners were locals or not ... Might have a few drinks and then start throwing the money around ... probably a bit of 'I'm a miner and I'm well paid and I'm better than you' sort of attitude.

Well-paid workers in the resources sector exaggerate the 'patchwork economy' in frontline towns, especially where owners of severely drought affected rural properties were experiencing financial pressures including losses of crops and livestock, even farms. Where the agricultural industry has been historically dominant, successful men of the 'squattocracy' and their family members have traditionally been associated with social, political and economic supremacy. Some farmers of the 'old' establishment harboured resentment towards the large numbers of mining industry workers in their communities. When asked if this factor upset those used to having influence and power, the response from the hotel manager was: 'It does a bit, particularly with them being used to being in that situation themselves.'

Hostilities were not overtly apparent between the farmers and the resident miners in 'Bramely'. Residents, irrespective of their workplace or occupation, were aligned instead against the transient non-resident intruder. The tendency for social division was illustrated by a long-term Denebola professional in this way: 'It's not too bad here because a lot of the farmers' families have gone off to be miners ... but you go to some places ... in [Town Z] and the mine is fairly new so it's a 'them-us' mentality'.

Although some locals were tolerant towards individual NRWs and expressed empathy for their lifestyle difficulties, FIFOs/DIDOs as a group were less acceptable, especially those staying in work camps who were held responsible for many of the adverse impacts affecting frontline communities.

Figure 7.2   Bowen Basin Work Camps Vary in their Levels of Amenity

'Bramely's' Work Camp Culture and Community Change

Patterns of life peculiar to those accommodated in work camps and other single persons quarters do not fit well with those idealised in rural Australia where residents lead modest lifestyles and have more time for family life and community building, reflected in rates of social capital, volunteering and church attendance (Carrington, McIntosh and Scott 2010). Most NRWs in Serveton lived in work camps where general conduct was regarded as 'fairly well controlled'. Instant dismissal can follow identified unacceptable behaviour. The alternative of expulsion from the camp can be tantamount to dismissal because affordable local accommodation is generally limited and thus extremely difficult to source. One professional living in Denebola was aware that:

> They've been fairly well told that if they misbehave, they're out. Not only are they out, the company risks the contract. One company that I know of has lost its contract due to misbehaviour ... brawling in a pub or something. Basically the contractors got told: that's it; you're out.

Camps are highly masculinised places: 85-90 percent of the transient population which uses them is male. If couples are employed, they must have separate rooms. One female professional employed by a Denebola mine was told it was 'not appropriate' for her to stay in the camp and was, instead, accommodated in a share house within the town. Camps in this study location usually operated a 'dry mess'; that is, they did not sell alcohol. Nevertheless, hotel bars were readily accessible to the extent that, according to one professional employed at a mine, the camp 'might as well be a wet mess; they can have alcohol in their rooms; they just bring it back'. Ease of access to alcohol by NRWs and its link with violent behaviour was a recognised problem not only by this employee but also by a local police officer:

> They're wet camps in that there's alcohol permitted to be there but they don't sell alcohol so people either have their private drinks in their rooms or they go to a hotel or licenced premise and they come back intoxicated. That's when we have the problems ...

Alcohol is not the sole stimulant for troublesome or abusive behaviour. A camp administrator thought that the long shifts influenced unacceptable behaviour in the camp's canteen:

> They can get abusive and tired ... We tend to see them at the worst times. Like, they're tired after a 12-hour shift and they're hungry as well and they haven't eaten yet and if there's not something there that they want, they'll tend to lash out a little bit.

Limited scope for alternative leisure and recreation within the camps plays a role in alcohol abuse. One ex-NRW talked about his down-time at a camp: 'I'd sorta get half-hooked on TV, go and talk; other people went and drank too much even though they restricted the hours, everyone just came home and drank.'

There are obvious negative implications for NRWs as well as for frontline communities. This experience was related by a senior police officer from Serveton:

> I've a future son-in-law who's been doing a bit of that and he reckons you can't sleep there ... Some of these fellas are obviously married with their wives [elsewhere]; flying back at the weekends, that type of thing, you know. Yeah, one fella there was in charge of the place; he couldn't wait to get out of this bloody hole, you know, and he'd just broken up with his wife because he was away and lost half his stuff ... I started thinking, these poor buggers, what sort of life have they had.

Not only are NRWs away from their homes for long periods but they are also subject to workplace rosters which appear to be detrimental to healthy lifestyles. These workforce arrangements affect resident workers as well as NRWs in the industry.

Workplace Roster Regimes – Work Hard, Play Hard

Changing workforce practices have manifold implications for the social organisation of everyday life in frontline communities. For example, the 12-hour shifts and irregular work patterns prevent participation in many community and family activities. These rosters were linked with a range of troublesome, violent incidents. For instance, because pub patronage tied in with workers' rosters, the 'best' trading days for publicans (or worst in terms of anti-social behaviour for town residents) were not restricted to one or two nights a week but instead coincided with the end of work cycles which could occur on any day or night of the week. Rosters can have workers switching between day- and night- shifts within work and leave cycles of varying length, all within a single roster. Pub violence was also usual when workers had single days off. When workers change from day shifts to night shifts, they have a 24-hour period without work, colloquially referred to as 'pyjama days' although opportunities to sleep were often substituted by drinking. A police officer remarked:

> We do have problems on pyjama days in town or if there's heavy rain – if anyone can remember what that is – out there and they actually stop production, for some of the contractors, we'll have a problem in the pub. They'll frequent the pub and more often than not, there's some violence.

A workplace culture condoning alcohol abuse can be influenced through the process of 'organisational osmosis', the seemingly effortless adoption of ideas, values and experiences (Gibson and Papa 2000). Workers' belief that they have earned the right to 'nights on the bender' when their work cycles finished was presented as the norm by community representatives; for example:

I think it's just the long work hours. So a lot of these contractors are on 12-hours a day, six days a week, with one day off. So they knock off, go to the pub and get blind. (Economic Development Manager)

Partying seems to be pretty common with them ... work hard and play hard. (Local government councillor)

Some of those we interviewed volunteered that FIFO/DIDO workforce practices promote a drinking and drug-taking culture. According to one police sergeant:

If you live locally, it's fine but if you don't, it encourages binge behaviour. If they've got a stable environment at home it might work but a lot of the time, it's work hard, play hard; work hard, play hard; work hard, play hard.

However, it is not only the NRWs who have difficulties with the long shifts and work cycles, as described by this experienced Registered Nurse in Denebola: 'They have their days off, it's very hard for them to pack up and go somewhere on their days off because their children are at school – the ones that live in town. And so there's a lot of drinking'.

Ill-considered and poorly coordinated rosters for NRWs on shutdowns – extended periods for scheduled plant and/or equipment maintenance – sometimes present local communities with unnecessary dilemmas and the potential for social disorder, at times with violent consequences. This increases demand on emergency services which deal with the fallout, as this senior ambulance officer explained:

You're now going to have these guys coming off six 12-hour shifts, very tired and then they land on a Friday night; they finish, straight into town, nightclubs and pubs and, you know, that's when the rate of violence went up there dramatically ... the workers might think it's great to have the Friday night off. The end result is more violence but that seems to be just a continuing thing, that contract work arrangement.

Incentives to reward work crews which record no or minimal workplace injuries can lead to hidden injuries not being treated or detected, with serious safety consequences. One offshoot of this is that when workers suffer injuries, they can feel pressure out of loyalty to hide the extent of their injury. An emergency services worker we interviewed was aware of one NRW trying to mask a severe back injury. Only when the NRW allegedly contemplated suicide using the morphine prescribed for pain relief was he sufficiently alarmed to call for help:

That frightened him. When I got to him, he was a real mess, in tears; just a broken man... Workcover actually wanted him to go on leave and have lost time injury but he said if he did that, the team would lose its bonus and he wouldn't be picked up for the next shutdown period by the company. I note that at that stage he went back to Brisbane to his family ...

Industry rosters normally translate as limited down-time during work cycles. The more continuous days of 12-hour shifts, the greater the potential for fatigue. Anecdotal accounts by some respondents linked extended work rosters with perceived behavioural changes in family members including increased agitation and aggression. Hostility and aggressive behaviour are linked with stress in the workplace (DiMartino 2003, Pocock 2001, Zitzmann and Nieschlag 2001). Fatigue associated with rosters and DIDOs has, nevertheless, been recognised by the mining industry as a significant health and safety issue which some companies attempt to address through education and training programs for workers (Petkova *et al.* 2009). Certainly expensive, powerful, 'hotted up' cars were closely linked with the high wages offered to workers in the mining industry. Connections have been made between DIDOs and the increasing number of road fatalities in the region (Petkova *et al.* 2009). One local businessman remarked:

> Here it is, you've got this mining road for I don't know how far it is and you've got these double white lines and hills and there's blind corners and I reckon every 50 metres there was a cross on the road: one dead, one dead, serious accident. Unbelievable! I think we stopped counting around 80 or something … Um, yeah, and the crosses are there to prove that.

The threats from this form of 'FIFO-fatigue' are widely dispersed and include workers who fly out and then drive home for the final leg of their journey. Not only are in-transit NRWs implicated but also the wider public and frontline emergency services personnel. This especially applies the Bowen Basin where, in July 2011, 77 percent of NRWs were DIDOs (OESR 2012). Workplace rosters, recruitment policies aimed at NRWs, shortage of affordable accommodation in

Figure 7.3   Vehicles Using Narrow Rural Roads in the Bowen Basin Compete for Space

proximity to workplaces, and inadequate road infrastructure all contribute to high road fatalities in the region. Until these factors are addressed by industry and governments, increased traffic densities and serious traffic accidents and fatalities will most likely remain a reality. These are among the challenges forced upon frontline rural communities by resource sector workplace practices.

Implications for Frontline Communities and Workers

The substantial impact of rapidly evolving mining industry regimes on the social ecology has suggested growing concern that typically few benefits accrue to those communities 'hosting' large numbers of NRWs (see Cleary 2012, 2011, Carrington and Hogg 2011, Murray and Peetz 2010, Haslam McKenzie *et al.* 2008). Most likely some businesses have increased levels of demand for their products or services, especially during project development stages. Expenditure on alcohol, on fuel and on some consumable items, mainly food, also brings ongoing benefits local economies. These perceptions are consistent with a 2007 survey in one Bowen Basin town which found highest expenditure by NRWs in work camps was on, firstly, alcohol ($52.66 per week) followed by fuel ($34.62 per week) and then food ($24.43 per week) (Rolfe *et al.* 2007). Nevertheless, in a more recent survey of Bowen Basin communities, the overwhelming response to perceived impacts of NRWs on the local economy, employment and infrastructure was negative (Carrington and Pereira 2011). Instead of reaping rewards, they can be lumbered with new burdens for which they are ill-prepared. Work shifts and rosters limit resident worker and family member involvement in and ongoing commitment to recreational and community activities and suppress NRW participation. A local government councillor's experiences describe just some of the associated difficulties:

> Thing is, you can't get people to come and work on your committees. They won't take it on because it clashes with their roster. And then the next thing you have, the president might be there but the secretary, he's on the next bloody shift so he's buggered too ... It's been happening since the seven day roster came in; it's a big impact.

In fact, in our study location, some football teams and even church congregations from geographically dispersed towns within the LGA had been amalgamated so that some (diminished) recreational and spiritual opportunities could still be provided. The generated obligation for additional travel on local road networks was thought by locals to increase the risk of traffic accidents. A grandmother and long-term local resident explained:

> The football used to be a big thing here but they've got teams up to about Under 10 but that's about all they've got. And for the older boys that want to play

football, there's not enough to make a team and so they join with [Canopus]
and make a combined team and they've got to travel over to [Canopus]. So that
then puts them at risk because there's kangaroos on the road and, it's just, I don't
know.

Some sporting clubs had resorted to paying for secretarial and treasury services
due to lack of volunteers. A local businessman elaborated on some reasons for this:

But once, here, all the parents would come to the cricket or football or tennis and
afterwards set off home with their kids. Now they give them a couple of bucks to
pay for their game … [parents] just won't take any positions on because of the
fact that a lot of kids are just left there and you're responsible for 'em as a team,
as a manager or president; if the kid's left there, you're responsible.

The erosion of community volunteering and wellbeing, the additional load on
local services and the stretch on infrastructure along with soaring housing costs
and other local costs of living (Carrington *et al.* 2012, Haslam McKenzie *et al.*
2008). Furthermore, 'fly-over' economic effects and a declining resident workforce
undermine economic diversification and long-term sustainable community
development (Carrington, Hogg and McIntosh 2011, Gallegos 2005) and foster the
tensions between resident insiders and the NRW outsiders (Cleary 2012, 2011), some
of which manifests as alcohol fuelled male-on-male violence (Carrington, McIntosh
and Scott 2010). When the local rumour mill pointed to increased activity of drugs
in the community, increased levels of violence also became apparent. Furthermore,
imbalance of the sexes meant that jealousy, especially when fuelled by substance
abuse, triggered hostilities, notably pub(lic) violence.

Mining employees readily acknowledged that sexual harassment 'just went
with the territory' of these male-dominated workplaces. This type of harassment
also carried over into the work camps. Conversely, accounts of specific instances
of sexual violence or rape were limited to well-publicised cases or based on
hearsay. Similarly, there was widespread ignorance about whether homosexuality,
homophobic crimes and even domestic violence were 'issues' in 'Bramely'. A
Community Centre Manager responsible for the provision of domestic and family
violence counselling observed, however, that in her dealings: 'People say there is
no domestic violence but we get a lot of calls from Denebola'. Another support
worker thought that 'perhaps there is more of a prevalence of people … putting
on a rough exterior, you know: I'm a hard working, hard drinking, hard fighting
mining man'.

These matters might have remained 'under the radar' were it not for the scale
and haste of new project developments which have forced acknowledgement that
some communities are bearing the brunt of Australia's rush to mine. Recognising
the full extent of NRW numbers would not only determine fairer allocations of
government funding but is also crucial for weighing impacts on frontline rural
communities. The 'switching' of NRWs, mostly men, between work and leave

cycles means that transients within the population are continually presenting as a different 'mix'. Using the example of Moranbah in the Bowen Basin, the 3,550 NRWs counted in a survey during July 2011 (estimated resident population at that time of 8,980) (OESR 2012) would be 3,550 different persons in the next work cycle if even-time rosters applied. For Western Australia's Pilbara Region, rosters of two weeks on to one off would have translated as an estimated 57,510 different individuals in 2010 over one roster period.

In these frontline settings, the efficiency of local services and their ability to cope are also compromised by rotating numbers of NRWs. Arguably none are more challenged than local doctors. Large numbers of rotating workers also put pressure on local infrastructure, and conspicuously on hospitals and transport routes. Moreover, rivalry between local men who consider themselves the authentic bearers of masculinity for their area and NRWs can trigger aggressive behaviour (Carrington, McIntosh and Scott 2010). This includes competition over local women where the potential for conflict increases with greater numbers of transients which further amplifies gender imbalance. NRWs are the classic 'outsiders' (Elias and Scotson 1994, Becker 1966) because they exist outside the informal social controls that traditionally characterise rural culture (Scott, Carrington and McIntosh 2011). This has manifold consequences for the social organisation of everyday life and patterning of frontier cultural conflict where workforce practices associated with NRWs have become emblematic of the cultural upheaval, disorder, destruction and loss being experienced by residents, where mining is regarded as the root cause (Carrington *et al*. 2012). In such contexts, the deviance of NRWs can be exaggerated. Furthermore, outsiders can be blamed for increasing crime levels in rural communities (Cohen 2001), deflecting attention from the social dynamics of a potentially wider culture of violence (Carrington and Scott, 2008). Resident feelings of safety within their communities become heightened, intensifying the 'us' and 'them' barriers between NRWs and locals (Carrington, McIntosh and Scott 2010). Concern about 'stranger' FIFOs/ DIDOs is further strengthened by the greater workforce turnover rate for NRWs, up to double that of resident employees (NRSET 2010). Residents see themselves as having long-term commitments to their communities but disproportionately bearing the costs of resource sector workforce practices.

Commuting workforces, long shifts and extended work cycles have become the preferred workforce practice of the resources sector irrespective of existing settlement patterns in mine locations. Well-paid employment opportunities appeal to many workers and their families. NRWs are seen to benefit substantially under these post-industrial mining regimes. But do they? There are handsome economic rewards for FIFOs/DIDOs but one could hardly devise work routines more hostile to sustainable family and community life. Strains on personal relationships due to the systematic workforce practices have been widely acknowledged by professional counsellors. Routine separation from family, support networks, informal social controls and a sense of belonging to a community can have serious negative impacts on the wellbeing of NRWs and their families, among them suicide, family breakdown and violence, alcohol and substance abuse, and fatigue related deaths and injuries (Ranford and Willcocks 2011, Lozeva and Martinova 2010, Carter and Kaczmarek 2009, Piortta 2009, Taylor and Simmonds 2009, Kaczmarek and Sibbel 2008, Jefferson and Preston 2008).

Conclusion

There is no expectation of 'new towns' being built for very remote or short mine-life projects. Nor can a resident workforce be justified for short-term project construction and for out-sourced maintenance. Increasingly, however, workers are losing freedom to choose where they live due to unavailability of job opportunities and unaffordable family housing in frontline communities. The presence of large numbers of FIFOs/DIDOs has already transformed many of these communities as illustrated by our 'Bramely' case study. Adverse impacts on social, human, institutional, economic and environmental capital are intensified by the cumulative effects of multiple projects within those regions which are the focus of extensive resource sector operations. Benefits are not so readily identified but clearly some businesses such as hotels and service stations profit with some flow-on employment opportunities. Extraction, processing and export from operational projects are expected to continue for decades to come, well after impetus from the current rush to develop new ones has flagged. There are compelling reasons for addressing challenging mining industry workforce practices and their impacts on frontline rural communities in Australia.

Acknowledgement

Our research was funded under the Australian Research Council Discovery Project scheme (DP0878476). Other team members were Professor John Scott and Associate Professor Russell Hogg.

References

ABS (Australian Bureau of Statistics), 2008, Towns of the mineral boom. *4102.0 – Australian Social Trends, 2008*, Canberra: ABS.

ABS (Australian Bureau of Statistics), 2012a, *1301.0 – Year Book Australia, 2012*, Canberra: ABS.

ABS (Australian Bureau of Statistics), 2012b, 6291.0.55.003 Labour Force, Australia, Detailed, Quarterly, May 2012, Canberra: ABS.

Australian Parliament, 2011, *Inquiry into the Use of 'Fly-in, Fly-out' (FIFO) Workforce Practices in Regional Australia*. [Online]. Available at http://www.

aph.gov.au/Parliamentary_Business/Committees/House_of_Representatives_
Committees?url=/ra/fifodido/tor.htm [accessed: 24 August 2011].

Becker, H., 1966., *Outsiders: Studies in the Sociology of Deviance*, New York:
The Free Press.

BREE (Bureau of Resources and Energy Economics), 2012, *Mining Industry
Major Projects, April 2012*, Canberra: BREE.

Carrington, K. and Hogg, R., 2011, 'Benefits and Burdens of the Mining Boom for
Rural Communities, *Human Rights Defender*, 20(2): 9-12.

Carrington, K., Hogg, R. and McIntosh, A., 2011, 'The Resource Boom's
Underbelly: The Criminological Impact of Mining Development', *Australian
and New Zealand Journal of Criminology*, 44(3): 335-354.

Carrington, K., Hogg, R., McIntosh, A. and Scott, J., 2012, 'Crime Talk, FIFO
Workers and Cultural Conflict on the Mining Boom Frontier', *Australian
Humanities Review*, Tourism and Mining Special Edition 53(November).
[Online]. Available at http://www.australianhumanitiesreview.org/archive/
Issue-November-2012/carrington_etal.html [accessed 6 December 2012].

Carrington, K., Hogg, R., McIntosh, A. and Scott, J., 2011, Submission No. 95,
House of Representatives Standing Committee on Regional Australia Inquiry
into the Use of 'Fly-in, Fly-out' (FIFO) Workforce Practices in Regional
Australia. [Online]. Available at http://www.aph.gov.au/Parliamentary_
Business/Committees/House_of_Representatives_Committees?url=/ra/
fifodido/subs/Sub95.pdf [accessed 9 Oct 2011].

Carrington, K., McIntosh, A. and Scott, J., 2010., 'Globalization, Frontier
Masculinities and Violence: Booze, Blokes and Brawls', *British Journal of
Criminology*, 50(3): 393-413.

Carrington, K. and Pereira, M., 2011, 'Assessing the Social Impacts of the
Resources Boom on Rural Communities', *Rural Society*, 21(1): 2-20.

Carrington, K. and Scott, J., 2008, 'Masculinity, Rurality and Violence', *British
Journal of Criminology*, 48(3): 641-666.

Carter, T. and Kaczmarek, E., 2009, 'An Exploration of Generation Y's Experiences
of Offshore Fly-in/Fly-out Employment', *The Australian Community
Psychologist*, 21(2): 52-66.

Cleary, P., 2012, *Minefield: The Dark Side of Australia's Resources Rush*,
Collingwood, Victoria: Black Inc.

Cleary, P., 2011, *Too Much Luck: The Mining Boom and Australia's Future*,
Collingwood, Victoria: Black Inc.

Cohen, S., 2001, *States of Denial*, Cambridge: Polity Press.

Deloitte Access Economics, 2011, *Queensland Resource Sector State Growth
Outlook Study*, Brisbane: Queensland Resources Council.

DiMartino, V., 2003, *Relationship of Work Stress and Workplace Violence in the
Health Sector*, Joint Programme on Workplace Violence in the Health Sector,
Geneva. Available at http://www.world-psi.org/Content/ContentGroups/
English7/Sectors/Health1/Safety_at_work_webpage/Publications16/EN_
relationship_of_workstress_workplace_violence.pdf [accessed 13 May 2011].

Elias, N. and Scotson, J., 1994, *The Established and the Outsiders: A Sociological Enquiry into Community Problems*, London: Sage.

Gibson, M.K. and Papa, M.J., 2000, 'The Mud, the Blood, and the Beer Guys: Organizational Osmosis in Blue-collar Work Groups', *Journal of Applied Communication Research*, 28(1): 68-88.

Haslam McKenzie, F., 2010, *Mining and Community Assessment in Western Australia*, 2010 Mining and Community Research Forum, Mackay, Central Queensland University, 16 June.

Haslam McKenzie, F., Brereton, D., Birdsall-Jones, C., Phillips, R. and Rowley, S., 2008, *A Review of the Contextual Issues Regarding Housing Dynamics in Resource Boom Towns*, Perth: AHURI Positioning Paper No. 105.

Houghton, D.S., 1993, 'Long-distance Commuting: A New Approach to Mining in Australia', *The Geographical Journal*, 159(3): 281-290.

Jefferson, T. and Preston, A., 2008, 'Western Australia's Boom Economy: Insights from Three Studies', *Journal of Australian Political Economy*, 61: 181-200.

Gallegos, D., 2005, '"Aeroplanes Always Come Back": Fly-in Fly-out Employment: Managing the Parenting Transitions', Perth: Centre for Social and Community Research, Murdoch University.

Kaczmarek, E.A. and Sibbel, A.M., 2008, 'The Psychosocial Well-being of Children from Australian Military and Fly-in/Fly-out (FIFO) Mining Families', *Community, Work & Family*, 11(3): 297-312.

Langton, M., 2010, 'The Resource Curse', *Griffith Review*, 28: 47-63.

Lozeva, S. and Martinova, D., 2008, *Gender Aspects of Mining: Western Australian Experience*, Perth: Curtin University of Technology.

Mayes, R., 2008, *Living the Resources Boom: Towards Sustainable Rural Communities*, Perth: Curtin University of Technology.

MISHC (Minerals Industry Safety and Health Centre), 2012, *Industry Profiles by Country: Australia*. [Online]. Available at http://www.mirmgate.com/index.php?gate=appgate&pageId=document&docId=2&docItem=5 [accessed 24 September 2012].

Murray, G. and Peetz, D.R., 2010, *Women of the Coal Rushes*, Sydney: UNSW Press.

NRSET (National Resources Sector Employment Taskforce), 2011, *Resourcing the Future*, Canberra: Commonwealth of Australia.

OESR (Office of Economic and Statistical Research), 2012, *Bowen and Galilee Basins Population Report, 2011*, Brisbane: Queensland Government.

Petkova, V., Lockie, S., Rolfe, J. and Ivanova, G., 2009, 'Mining Developments and Social Impacts on Communities: Bowen Basin Case Studies', *Rural Society*, 19(3): 211-228.

Pini, B., McDonald, P. and Mayes, R., 2012, 'Class Contentions and Australia's Resource Boom: The Emergence of the "Cashed-up Bogan"', *Sociology*, 46(1): 142-158.

Pirotta, J., 2009, 'An Exploration of the Experiences of Women who FIFO', *The Australian Community Psychologist*, 21(2): 37-52.

Pocock, B., 2001, *The Effect of Long Hours on Family and Community Life: A Survey of Existing Literature*, Brisbane, Queensland: Centre for Labour Research, Adelaide University for the Queensland Department of Industrial Relations.

Ranford, A. and Willcocks, A., 2011, *The Survival Guide for Mining Families*, Mining Family Matters, www.miningfm.com.au.

Rolfe, J., Petkova, V., Lockie, S. and Ivanova, G., 2007, *Mining Impacts and the Development of he Moranbah Township (Research Report No. 7)*, Mackay, Queensland: Centre for Environmental Management, Central Queensland University.

Scott, J., Carrington, K. and McIntosh, A., 2011, 'Established-outsider Relations and Fear of Crime in Rural Towns', *Sociologia Ruralis*, 52(2): 147-169.

Storey, K., 2001, 'Fly-in, Fly-out and Fly-over: Mining and Regional Development in Western Australia', *Australian Geographer*, 32(2): 133-48.

Storey, K., 2010, 'Fly-in/Fly-out: Implications for Community Sustainability', *Sustainability*, 2: 1161-81.

Taylor, J. and Simmonds, J., 2009, 'Family Stress and Coping in the Fly-in Fly-out Workforce', *The Australian Community Psychologist*, 21(2): 23-36.

URS, 2012, *Workforce Accommodation Arrangements in the Queensland Resources Sector*, Brisbane: Queensland Resources Council.

WACME (Western Australian Chamber of Minerals and Energy), 2005, *Fly in/Fly out: A Sustainability Perspective*, Perth: WACME.

Watts, J., 2004, *Best of Both Worlds? Seeking a Sustainable Regional Employment Solution to Fly-in Fly-out Operations in the Pilbara*, Karratha: Pilbara Regional Council.

Zitzmann, M. and Nieschlag, E., 2001, 'Testosterone Levels in Healthy Men and the Relation to Behavioural and Physical Characteristics: Facts and Constructs', *European Journal of Endocrinology*, 144: 183-197.

Chapter 8

Trajectories of Change in Rural Landscapes: The End of the Mixed Farm?

Fiona McKenzie

Introduction

Agriculture in many developed countries has undergone a transformation –
becoming both increasingly intensified and specialised. Accompanying this
trend has been a decrease in mixed farming systems and the ongoing substitution
of labour for capital. This chapter explores these trends in the Australian context
where many competing forces are shaping the agricultural landscape. While
recognising that economic, technological and food security factors are key forces,
the chapter focuses on the role of knowledge intensity as a driver of farmers'
decisions to specialise in both commodity production and land use. The growing
knowledge intensity of agriculture, accompanied by a concentration in human
capital, is putting increased pressure on farm managers to maximise investments

Figure 8.1   Mixed Farm in Central West NSW – A Relic of the Past?

in knowledge and skills by specialising. This specialisation is resulting in a reduction in the number of land uses on individual farms, in turn placing both the traditional mixed-farming system at threat as well as the ecological diversity of agricultural landscapes. This is a new trend as the agricultural industry rapidly undergoes change. Better recognition of the important role of human capital is an urgent challenge, not just for farmers or for sustainability, but for the wider industry as a whole.

Linking Specialisation and Knowledge Intensity in Agriculture

Specialisation is the concentration on a limited number of products or land uses, which can lead to less diverse cropping and/or livestock patterns. Specialisation is often, but not always, associated with intensification (the increasing of productivity through the greater use of energy, capital, chemicals and machines). Agricultural specialisation is occurring in most developed countries. For example Western Europe has been experiencing the intensification, separation and specialisation of agricultural and livestock systems since the middle of the twentieth century (Kirkegaard *et al.* 2011). North America has followed a similar trend where, in places like Iowa, farms have become increasingly specialised grain or livestock producers since the 1950s (Brown and Schulte 2011). Until recently, Australia did not follow this trajectory. Farmers traditionally retained a mix of cropping and livestock enterprises ('mixed farming') as a risk management strategy that gave them the flexibility to respond to climate and market variations. However, in the past two decades, Australia has begun to see a shift. In the rangelands and semi-arid regions where livestock has always dominated, mixed farming continues, but in the higher rainfall areas of the eastern and western wheat/sheep belts (see Figure 8.2), there has already been a swing away from traditional mixed crop-livestock systems towards either crop or livestock systems. From 1992-93 to 2009-10, the area planted to crops (excluding pastures and grasses, and hay) increased by 50 percent, while grazing area decreased by 6 percent (Lesslie *et al.* 2011).

There are a number of reasons why farmers may choose to specialise, with economic explanations predominating. For one thing, specialisation has the potential to allow farmers to maximise product quality and volume through efficiencies of scale (Chavas 2008). Since the 1990s, greater profits have been available from cropping than from either sheep or beef cattle production (Kirkegaard *et al.* 2011). Specialisation is also a rational response to declining terms of trade and the need to improve financial returns through the increased concentrations of inputs such as nutrients, water, energy and management effort (Lesslie *et al.* 2011). Although these are all important justifications, they are not the only reasons for specialisation. Other less tangible factors can influence decisions to specialise – factors such as human capital: the combination of

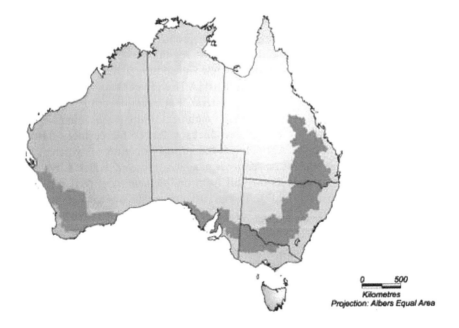

Figure 8.2   Eastern and Western Wheat-Sheep Zones

Source: ANRA 2001

knowledge and skills (Schultz 1961). The important role of human capital is well recognised in other industries, such as manufacturing and the services sector. While there have been studies on the impact of human capital on farm productivity, little attention has been paid to the role of human capital as a driver of specialisation and hence of land use change in agriculture.

Why are knowledge and skills so significant? Knowledge is not the same as information. It is not simply additive, like data collection, nor is it easily abstracted from its context (Wolf 2008). It can include perceptions, implicit understandings, unconscious motivations and behavioural habits: knowledge cannot simply be given to someone else (Breschi and Malerba 2001). It has to pass through a filter of perceptions and interpretations and it takes time and effort to build (Midgley 2000). Traditionally, farmers' knowledge has been characterised as local, tacit and informal, formed through practical advice from other individuals plus a farmer's own practical experiences (Oreszczyn *et al.* 2010). Today, farmer's knowledge could be classed as global as well as local, formal and scientific as well as tacit and practical. Agriculture is very different now from how it was even a decade ago.

The modern agro-industrial system is frequently characterised as intensive in terms of external inputs and energy use. What is less well known is that modern agriculture is also becoming more knowledge intensive. Agricultural systems and their management have become increasingly complex, underpinned by expensive capital investments, changing production technologies, volatile markets, social challenges and increased regulation. This has increased management complexity and has placed new and additional demand on farmers' existing knowledge and skills (Kingwell 2011). This greater knowledge intensity is true of both cropping and livestock systems. For example, cropping systems now require specialised skills to manage technical, biophysical, financial and marketing components of the business (Jackson 2010). Increasingly farmers are adopting Conservation Agriculture, the primary feature of which is the maintenance of a permanent or semi-permanent soil cover, either a live crop or dead mulch, which protects the soil from the elements and feed soil biota (Knowler and Bradshaw 2007). It can involve reducing soil compaction through 'controlled traffic farming' or the use of predetermined tracks or 'tram lines' guided by a Global Positioning System (GPS). Many farmers now also manage variable land and soil conditions on a micro-scale through 'precision agriculture', which involves the use of high resolution spatial data and analysis, and tailoring inputs and management accordingly (CSIRO 2011). Even planting a new crop is not simple. Activities such as the management and application of fungicides, herbicides and pesticides and management of subsoil water content, all require detailed knowledge, skills and experience. Alongside these technologically complex processes are post-harvest elements of cropping, namely bulk handling, storage, transport and marketing. Variable returns depend on market destination, choice of terminal or port for grain delivery, forward selling against fluctuating volumes and exchange rates. The timing of sales and contracts plays a key role in an industry that is globally connected and subject to dramatic fluctuations in global commodity prices.

The livestock sector is often viewed as the less sophisticated cousin of cropping, but advances have also occurred in the sector over recent decades, increasing its complexity and the knowledge required by graziers. To transport and sell livestock in Australia, farmers must comply with the National Livestock Identification System (NLIS), which became mandatory in 2006. The NLIS enables individual animals (cattle, sheep and goats) to be tracked from birth to death for purposes of biosecurity, meat safety and market access. Farmers are required to fit each animal with an identity tag and record livestock movements and transactions on the NLIS database. Online auctions, vendor declarations, weighing livestock to cater to weight-specific orders, pregnancy scanning, chemical record management and paddock rotations all add to the complexity of running a contemporary livestock enterprise. Adequate grass for grazing is no longer simply a matter of waiting for rain. Farmers implement different grazing strategies to restore perennial plant density, suppress annual grasses and increase soil seed reserves. To better predict the performance of an animal based on its genetics, an 'estimated breeding value' can now be calculated using software.

Such examples demonstrate the increasing complexity of agricultural systems and the greater investments required by farmers in human capital (knowledge and skills). This has implications for land use change because investments in human capital create incentives to specialise, since there are fixed costs to knowledge acquisition, regardless of how that knowledge may later be used. To maximise the rate of return on investment, specialised skills and knowledge must be used as intensively as possible. The greater the investment, the greater the incentive.

Evidence from New South Wales

The following section draws examples from research conducted across the wheat/ sheep belt in New South Wales (NSW). This research – the NSW study – investigated farmers' experiences with on-farm innovation, decision making and knowledge management, and was conducted in mid-2009. Purposeful sampling targeted 33 farmers who had a record of innovative land management on 22 broadacre dryland family farms. A relatively small number of farmers were intentionally chosen, facilitating in-depth research across a geographically dispersed area.

"GP versus Your Specialist"

The NSW study demonstrated the impact of knowledge intensity on farm management decisions. Some 16 of the 22 farms were mixed farms in mid-2009; however 21 of the 22 had been mixed farms in the past, with 5 having changed from mixed farms to specialist cropping enterprises since 1993. That trend is consistent with a wider trend in Australia, where cropping has become increasingly important in many farming enterprises, often totally replacing livestock. From an economic point of view, the shift towards cropping in Australia is usually explained by the decline of the sheep industry. Sheep numbers peaked in 1990, followed soon after by the collapse of the wool stockpile in 1991. A crisis in the industry led to the introduction of the Flock Reduction Program, which paid farmers to destroy over 20 million sheep (Rudwick and Turnbull 1993). Farmer #4 had to destroy 2500 sheep, which was the "worst job of my life". On top of the collapse in market prices, prolonged drought saw breeding herds of sheep and cattle being sold off and eventually replaced with trading stock, such as steers and wethers, which are owned for a much shorter period of time and require less emotional or economic investment than breeding stock. However, after decades of decline, the livestock sector may be experiencing a renaissance. While productivity growth has been higher for cropping specialists than for mixed crop-livestock farmers or livestock specialists for almost 30 years, productivity of livestock specialists has recently begun increasing (Jackson 2010, Mullen 2007, Nossal *et al.* 2009). Beef specialists now achieve the same average performance level as the mixed crop-livestock industry (Nossal *et al.* 2009). Sheep, lamb and meat cattle numbers increased in 2010-11 while farm cash income for the sheep industry has returned to levels not seen since 1989-90.

From a knowledge point of view, it is not surprising then that when making a choice to specialise, farmers are being drawn to the cropping sector because of the greater investments being made in research and development. As Farmer #23 explained:

> Cropping – [is] where all the new technology is, where all the bigger toys are, where more money is, greater support infrastructure....... try and find the same level of information and everything else around that is run on the livestock point of view and it is just not there.

These investments have further facilitated crop specialisation through developments like nitrogen fertilisers, herbicides, fungicides and new 'break crops' which make continuous cropping possible where previously pests, diseases and nutrient deficiencies would have prevented this approach (Kirkegaard *et al.* 2011). In addition to knowledge availability, knowledge intensity played a role as well. For example, Farmer #2 felt that "it is timing – it is all time dependent" while Farmer #29 explained, "these days, more and more, it is the timing. Timing is the big thing as I understand it. Like, you have got to kick in and do things when they need to be done". Farmer #15 explained how they had moved from being a mixed farm to just a cropping enterprise because it meant they had "been able to get our timing much better on our cropping" whereas "cattle used to take 60 percent of our labour and make about 20 percent of the money". Timing and making the most of windows of opportunity provided by seasonal conditions are crucial in cropping. But getting timing right requires knowledge and skills, which is in turn time consuming and an incentive to specialise.

For Farmer #15, it didn't matter whether farmers chose to be graziers or croppers, but "if you are a grazier, then rather than do a bit of part time cropping, you're probably better off to be a straight out grazier". Farmer #10, who had also specialised, argued that "it is very unusual to see a really good cropping manager and a really good stock manager in the same person". Farmer #7 had a similar view. He likened it to the "GP versus your specialist. If I need brain surgery I don't want my GP to take it or open heart surgery – there is that level of professionalism now". Farmer #16 also found it was more effective to concentrate her efforts on one enterprise. She felt that farm management is "not going to be perfected in every area by one person ... you'd be too thinly spread".

These comments reveal that while diversifying can be a risk management strategy, it can also come at a cost. Human resources must be divided between more land uses, potentially undermining farm management, timing and business viability. In a mixed-farm, resources can be channelled away from the core business to support other secondary efforts (Grande 2011). In contrast to diversification, specialisation had allowed farmers to focus their knowledge and skills on fewer enterprises and to maximise their return on investments in human capital.

Although the agricultural industry has undergone profound changes, several farmers felt that society still held outdated views of what it was to be a farmer –

that the community at large still envisaged farming as a largely outdoors lifestyle involving the man on the land with his horse and his dog. In reality, as much time can be spent planning, researching, marketing and doing 'paper work' in the office as is spent outdoors working in the field. And many new machines and gadgets have replaced some of the more traditional roles on the farm. As Farmer #7 said of agriculture today:

> It is a sexy industry and no one knows about it...The only thing we need now is lasers and we've got satellites and robots and all these other things that every kid wants to work with and they all think we chew straw...

Concentration of Human Capital

Though mixed crop-livestock farming still dominates the major cropping zones in southern Australia, it looks very different from how it once was. In addition to being technologically complex, Australian agriculture is becoming an industry with fewer farmers and bigger farms. It therefore comes as no surprise that farmers share a sense of being too "thinly spread" (F#16). While the knowledge and skill requirements of agriculture have been increasing, the number of people working in the industry has been decreasing. The replacement of human labour with capital and machinery can increase economic efficiency and reduce the burden of some tasks. However, it also concentrates human capital – knowledge and skills – in fewer individuals. Australian agriculture employs fewer people every year; between 2001-2006, employment fell by fully 19 percent (ABS 2008). In the past, family members would have been called on to fill any gaps in labour, particularly during peak times like shearing and harvesting. To some degree, this is still the case. At a global scale, agriculture remains one of the few industries still largely based on a family business model, even in countries like the United States (Alston 2010, Deininger and Byerlee 2012). But these figures can be misleading. In Australia, the proportion of broadacre and dairy farms that are family owned remains constant at between 94-99 percent (ABS 2003, Pritchard *et al.* 2007). However, while the ratio of family to non-family farms remains the same, the number of family farms has actually decreased significantly – by 30 percent between 1986-2006 – involving the loss of approximately 42,000 farms (with smaller farms often being absorbed into larger holdings).

The remaining family farms are getting bigger, but have fewer family members staying on to work on the farm. This can be partly driven by an increased dependence upon off-farm income as the profitability of agriculture declines (Barr and Cary 2000). While off-farm income can have its advantages, again, it has a cost in terms of human capital and farm diversification, not least because non-family farm labour is less readily available than in previous generations (Pfiefer *et al.* 2009). In the NSW study, labour issues were raised by more than half the farmers. As wives have increasingly taken up work off-farm and children have decided not to return to the farm after finishing their education, farmers have had to perform more tasks on their own. Farmer #28 explained:

> I've got three kids and a wife, and unfortunately I don't think any of the kids are going to go on the land. The oldest one definitely won't. The youngest one definitely won't. The middle one is doing business agriculture, but not with a bent to come back on the farm but to be a trader or something along those lines.

With fewer hands available, farmers must work smarter in order to remain viable. Given family dynamics, Farmer #17 was also reconsidering his enterprise mix, being inclined to concentrate more on livestock:

> I'm looking to expand the stock area and cut back a bit on the cropping because look, as much as anything, I do the whole thing on my own. I run the whole farm on my own ... it's just a very big job. And my kids are getting a bit bigger and I'm just looking to try and wind back on that cropping a little bit and run more stock.

The reduction in family members working on farms not merely reduces labour availability but also represents the departure of trusted advisors. Family is one of the most important sources of information for farm managers and can play a crucial role in shared decision making. In the NSW study, farmers expressed a strong tendency to rely on family members and other farmers for advice. A 2007 study in Central West NSW similarly found that the most popular individuals for a farmer to seek information and advice from were family members, before other farmers, agribusinesses and agronomists (CSIRO 2007). As more family members work off-farm, their role as financial advisors, confidantes and business partners may be lost, potentially removing an important component of farm success.

Fewer family members remaining on-farm can also mean the loss of skilled labour. Australian agriculture suffers not only a labour shortage, but the absence of a large pool of specialised and skilled labour. Skilled labour tends to reside within the owner/manager. Beyond reductions in family labour, there were difficulties in finding on-going skilled labour needed for day-to-day farm management. With the increasing complexity of agriculture, even day-to-day tasks require significant knowledge and skills. Once employees were found, they then had to be trained in the use of advanced technologies, appropriate management practices such as controlled traffic farming and renumerated appropriately. In the NSW study, farmers were concerned that non-family employees were an insecure investment, since it required the input of significant time and effort to develop their knowledge and skills. There was concern that once an employee had been trained or "up-skilled", there was nothing to stop them leaving and finding work elsewhere. Training family members was a more secure investment. Farmer #4, who had pooled his funds with several

Figure 8.3   Shared Decisions are Important in Farm Management

other neighbouring farmers to jointly employ a skilled farmhand, told how his neighbour was having problems with hiring, particularly at peak times like crop planting:

> A neighbour over here, a mate of mine ... I've seen him do 76 hours straight. Where he should be able to get on the phone and say, I need three blokes to drive this machine, 8 hours shifts, go. Can't do it – can't find the workers.

The reliance on family members reveals another dimension to the impact of human capital on decision making – the current crisis in succession. On many Australian farms, the main adjustment to declining profitability has been the abandonment of expectations of intergenerational transfer. There is a lack of young people entering the industry – family members or otherwise (Barr 2004). Older farmers are deferring farm exit, have an increased dependence on off-farm income, and do not expect to transfer the farm to another generation (Barr and Cary 2000, Barr *et al.* 2005). This is an issue among farm families in other developed countries as well, such as the United States, which is also experiencing a delayed transfer of assets to the next generation (Richards and Bulkley 2007). The deferral of farm exit in response to a lack of perceived alternatives has contributed to the increase in the average age of the farm population (Barr and Cary 2000, Barr *et al.* 2005). However, being older doesn't necessarily mean being less successful or innovative. Within the Australian cropping sector, productivity is highest for farmers aged between 55 and 60 years (Zhao *et al*. 2009). This is in part due to the value of accumulated knowledge and experience that is embodied in these older farmers. However, as ageing farmers retire without successors, knowledge that has been accumulated over generations is potentially lost. As Farmer #15 said:

> And all this information is in that head, it's nowhere else. It isn't anywhere else. And that's the trouble. There are a lot of farmers out there with some really good skills but it just gets lost.

Stagnation in succession planning further serves to concentrate human capital in the sector, with cascading impacts on the broader community. In Western Australia, economic restructuring and depopulation are undermining both social capital and the viability of sporting clubs and organisations (Tonts 2005). Farmer #28 echoed these concerns for New South Wales, recalling how:

> When I left school, I was playing Aussie rules and we had rules, union and league. And we had soccer players that used to travel away. Now we have got a league team, no union team and no rules team. There's just not the people.

This concentration of human capital, of knowledge and skills, means that individual farmers have to make a choice. They can continue to build expertise and skills in more tasks, or concentrate their efforts and expertise on fewer subjects. Eventually they are going to have to prioritise. This can result in some things being done less well, or in their removal. Consequently, cropping specialists are more likely than mixed farmers to introduce new crop husbandry practices, cultivars and fertiliser/ weed/pest/disease management practices (Nossal and Lim 2011). Parallels exist in other countries. In Denmark, Finland, Germany and Greece, the majority (50- 60 percent) of farmers who adopted precision farming practices (such as yield mapping) were cropping specialists, rather than mixed farmers (Lawson *et al*. 2011). These specialists are in a better position to adopt advanced technologies than those who were balancing multiple land uses.

A 'Hardening' of the Line Between Cropping and Livestock

The growing demand on knowledge and skills required to optimise farm performance makes specialisation attractive, even with the potential economic risks it can bring. In the past, the balance between cropping and livestock on mixed-farming systems could be varied based on market and climate conditions. The ability to adjust this balance may no longer be as great, due to infrastructure decline. Where infrastructure for either cropping or livestock no longer exists, the reversibility of specialisation and the opportunity to (re)integrate crop-livestock systems is reduced (Hochman *et al*. 2012). For example, in the NSW study, several

**Figure 8.4   A Stockpile of Water Troughs and Fencing Materials –
Removed after a Switch from Mixed Farming to Cropping**

farmers who had switched to cropping had already removed fences, water troughs and even shearing sheds and sheep yards.

New spatial arrangements on increasingly specialised farms also inhibit flexibility. In the NSW study, 10 of the 22 farms had undergone property redesign. In order to implement controlled traffic cropping, fence-lines were being adjusted to create bigger paddocks (to minimise the number of times a tractor or other GPS controlled machinery has to turn). Grazing paddocks were being made smaller as part of a rotational grazing system that sought to increase the planned movement of stock around a number of paddock cells. Rotational grazing is timed to coincide with the most nutritious stage of the plant growth cycle. Paddocks are rested between grazing events, allowing sufficient time for plants to regrow. The use of electric fences was widespread across the farms which ran livestock, suited to either sheep or cattle. The changing fence-lines reflected more than a lot of labour intensive work, revealing a broader conceptual shift in farm management. A hardening of the line between the two farming systems was occurring. Switching back and forth may not be as easy as it once was.

This 'hardening' of the division between land uses is ironically being driven by two practices which seek to increase the sustainability of farm management: rotational grazing and conservation agriculture. Conservation agriculture places a strong emphasis on the exclusion of livestock to avoid soil compaction. Rotational grazing emphasises the rotation of stock through small paddocks to allow the rejuvenation of perennial pastures. Both essentially require the exclusion of other enterprises. Farms become divided into two zones – the cropping and the livestock zones. The implication of such separation is that synergies or complementarity between the two systems are removed. Separation of the two enterprises can make the final step towards single land use specialisation easier. It also fundamentally changes the dynamics of the mixed farm. Complicating matters further is the fact that enterprise separation was also facilitated by availability of the very thing whose absence can also lead to specialisation – the next generation returning to the farm through successful farm succession. Where there was more than one family member or 'labour unit' responsible for farm management (often when a son or daughter returned to the farm), a division of labour was possible. Put simply, while labour shortages could mean specialisation across a whole farm, labour availability facilitated enterprise separation within a mixed farm.

Rediscovering the Mixed Farm

Do the above trends towards specialisation and enterprise separation mean the end of the mixed farm as we know it in Australia? Currently mixed-farms are on a trajectory towards distinct zones for cropping and livestock enterprises, either as separate entities within a farm or a single land use enterprises. However, a counter trend is also at work shaping the future trajectory of agricultural landscapes. A renewed awareness is emerging of the importance of mixed crop-livestock systems as a means of improving system diversity, nutrient cycling and other natural processes (Herrero *et al.* 2010, Villano *et al.* 2010). Such awareness may slow the separation of enterprises and instead cause re-integration.

Demand is being generated by farmers themselves for the re-integration of livestock and cropping systems. In the NSW study, re-integration was being driven by farmers' recognition (and sometimes rediscovery) of the role of livestock in nutrient cycling. While most farmers had adopted the prescriptions for conservation agriculture and had removed livestock from their cropland, several were beginning to question the cost of the separation. They either had plans, or had already begun, to reintroduce livestock into the cropping system or vice versa. They saw livestock as playing a key role in nutrient cycling and ongoing system health. Farmer#14 had decided to reintroduce some livestock into the cropping system because he believed that the soil biology could benefit from "a bit of manure in the system, a few hoofs in there". Meanwhile, Farmers #2, #3 and #33 were implementing a farmer-developed technique known as pasture cropping.

Pasture cropping involves the direct sowing (without ploughing) of annual cereal crops into living perennial pastures. It creates new synergies between grazing and cropping, as well as flexibility to move between the two. The crop increases biomass and soil health, whilst also providing forage for livestock and, if it yields enough, seed for harvest. Instead of bringing livestock into the cropping system to graze, this technique involves bringing cropping into the grazing system. As Farmer #33 said of livestock:

> You need them to cycle the grasses and promote biology. The crops actually help with the livestock too because they're providing high quality forage and they're providing another choice for animals to mix with dry grass. So they both help each other.

Gains from such integration may become more important as productivity growth slows down. Eventually mixed farms may recover the overlap between livestock and cropping, and create systems where there is greater balance between plant composition, animal density and nutrient cycling.

Specialisation or Sustainability?

These trends towards specialisation can confound alternative visions for sustainable agriculture. One example is the notion of 'multifunctionality' – a vision of agriculture as a source of social, economic and environmental sustainability served by multiple functions (with multiple land uses). In this vision, landscapes are on a trajectory where they evolve from a 'productivist' focus on producing food and fibre into 'post-productivist' places of resource management, biodiversity conservation, and the provision of ecosystem services. There is some sense of inevitability that this is the direction that agriculture will evolve, yet, there is nothing inevitable about such a transition. Rather than moving into more multifunctional or post-productivist modes of agriculture, production in many dryland agriculture areas is in fact becoming more intensified with fewer, not more, land uses.

Understanding knowledge intensity as a driver of specialisation also challenges assumptions that it can be a driver for sustainability. When farmers in the NSW study were asked to comment on an image of a 'future farm' – their responses were largely negative. It was one version of what a 'multifunctional' landscape might look like, with the farm divided into seven different land uses, with income generated through biodiversity credits, carbon offset credits, renewable electricity, certified sustainable timber, water credits, wheat and wool production. The image provoked a common response – that this 'future farm' was unrealistic because it comprised far too many land uses. It was best explained by Farmer #16 who felt that:

> To be run by one farmer is near impossible…to be able to cover all the other industries and to be up on all of that knowledge is, for one person, nearly impossible.

Farmer #15 expressed similar sentiments: "the first thing that springs to mind is the fact that you've got five or eight different enterprises there and I mean [what are] the chances of one person being able to be totally tech savvy at all of those?".

If knowledge intensity is driving farmers away from multiple land uses, what does this mean for calls for sustainable agricultural systems based on knowledge intensity rather than intensity of inputs such as energy, machinery and chemicals? Would UNCTAD's (2010) suggestion of transforming high-input industrial agricultural systems into knowledge intensive regenerative agricultural systems be possible? Knowledge intensity doesn't rule out sustainability (and may in fact be more benign than capital or energy intensity), but the rise in complexity accompanied by a decline in human capital and specialisation may be undermining ecological diversity within agricultural systems. This is true where specialisation results in more intensive land management practices, less diverse cropping/ livestock patterns, and negative environmental impacts. However, in other instances, specialisation, for example in rotational grazing for livestock, can lead to increased perennial and native grass species diversity and ecosystem health. The challenge is to ensure that where specialisation occurs, it isn't to the detriment of broader landscape sustainability. Where intensification must occur, it should be pursued along the lines of ecological intensification, defined as 'producing more food per unit resource use while minimising the impact of food production on the environment' (Hochman *et al.* 2013). A key opportunity for ecologically efficient intensification exists through the better integration of crop and livestock enterprises on mixed crop-livestock farms, so returning to the important role of mixed farming.

Conclusion

It is not yet clear which of the many competing forces shaping the agricultural landscape will dominate. Certainly the mixed farm of the future will look very different to farming systems of today, with complex and knowledge intensive management systems an integral part of the farm business. As pressures on mixed-farming accumulate, greater support for the knowledge and skills associated with mixed farming systems will be needed. If greater human capital was available, then more than one type of specialisation (a division of labour) might be possible within individual farms, maintaining land use diversity. Along these lines, mixed-farming itself could become a specialised skill rather than a compromise between enterprises. This would require acknowledging the human capital challenges facing mixed farming. It would also mean renewed focus on research and development challenges. Past efforts have largely focused on single system components or single enterprises. Australian industry organisations are not exempt, and have neglected the interaction between crops and livestock for years. What has resulted is a poor understanding of the complexities of inter-relationships between enterprises on mixed farms. Gaining a renewed appreciation of the benefits of multiple land uses for productivity will be an important means of not only moving towards a more sustainable agricultural sector, but ensuring that mixed farming systems survive. Only by better understanding the role of knowledge and skills will knowledge intensity become an asset instead of a burden, a driver of diverse agro-ecological landscapes and vibrant family farms.

References

ABS, 2003, 'Living Arrangements: Farming Families' – 4102.0 – *Australian Social Trends*, 2003, Canberra: Australian Bureau of Statistics (ABS).

ABS, 2008, *Agriculture in Focus: Farming Families*, Australia, 2006, Canberra: Australian Bureau of Statistics.

Alston, J.M., 2010, 'The Incidence of US Farm Programs', *The Economic Impact of Public Support to Agriculture: Studies. Productivity and Efficiency*, 7: 81-105.

ANRA (2001), 'Australian Agriculture Assessment 2001 – Profile of Australian Agriculture, Australian Natural Resources Atlas, Department of Sustainability, Environment, Water, Population and Communities'. Available from http://www.anra.gov.au/topics/soils/pubs/national/agriculture_profile.html.

Barr, N., 2004, 'Australian Census Analytic Program: The Micro-dynamics of Change in Australian Agriculture 1976-2001', Canberra: Australian Bureau of Statistics (ABS).

Barr, N. and Cary, J., 2000, *Influencing Improved Natural Resource Management on Farms: A Guide to Understanding Factors Influencing the Adoption of Sustainable Resource Practices*, Canberra: Bureau of Rural Sciences.

Barr, N., Karunaratne, K. and Wilkinson, R., 2005, *Australia's Farmers: Past, Present and Future*, Canberra: Land & Water Australia.

Breschi, S. and Malerba, F., 2001, 'The Geography of Innovation and Economic Clustering: Some Introductory Notes', *Industrial and Corporate Change*, 10: 817-833.

Brown, P.W. and Schulte, L.A., 2011, 'Agricultural Landscape Change (1937-2002) in Three Townships in Iowa, USA', *Landscape and Urban Planning*, 100: 202-12.

Chavas, J., 2008, 'On the Economics of Agricultural Production', *The Australian Journal of Agricultural and Resource Economics*, 52: 365-380.

CSIRO, 2007, *Partnerships and Understanding Towards Targeted Implementation: Identifying Factors Influencing Land Management Practices – An Overview*, Wembley, WA: CSIRO Land and Water.

CSIRO, 2011, 'Software tools for Precision Agriculture'. Available at: http://www.csiro.au/en/Outcomes/Food-and-Agriculture/Precision-Agriculture-Software.aspx. Accessed 31/5/12.

Deininger, K. and Byerlee, D., 2012, 'The Rise of Large Farms in Land Abundant Countries: Do They Have a Future?', *World Development*, 40: 701-14.

Grande, J., 2011, 'New Venture Creation in the Farm Sector – Critical Resources and Capabilities', *Journal of Rural Studies*, 27: 220-233.

Herrero, M., Thornton, P.K., Notenbaert, A.M., Wood, S., Msangi, S., Freeman, H.A., Bossio, D., Dixon, J., Peters, M., van der Steeg, J., Lynam, J., Parthasarathy Rao, P., Macmillan, S., Gerard, B., McDermott, J., Sere, C. and Rosegrant, M., 2010, 'Smart Investments in Sustainable Food Production: Revisiting Mixed Crop-livestock Systems', *Science*, 327: 822-825.

Hochman, Z., Carberry, P.S., Robertson, M.J., Gaydon, D.S., Bell, L.W. and McIntosh, P.C., 2013, 'Prospects for Ecological Intensification of Australian Agriculture', *European Journal of Agronomy*, 44: 109-123.

Jackson, T., 2010, *Harvesting Productivity: ABARE-GRDC Workshops on Grains Productivity Growth*, Canberra: ABARE.

Kingwell, R., 2011, 'Managing Complexity in Modern Farming', *Australian Journal of Agricultural and Resource Economics*, 55(1): 12-34.

Kirkegaard, J.A., Peoples, M.B., Angus, J.F. and Unkovich, M.J., 2011, 'Diversity and Evolution of Rainfed Farming Systems in Southern Australia', in P. Tow, I. Cooper, I. Partridge and C. Birch (eds), *Rainfed Farming Systems*, New York: Springer, 715-754.

Lawson, L.G., Pedersen, S.M., Sorensen, C.G., Pesonen, L., Fountas, S., Wener, A., Oudshoorn, F.W., Herold, L., Chatzinikos, T., Kirketerp, I.M. and Blackmore, S., 2011, 'A Four Nation Survey of Farm Information Management and Advanced Farming Systems: A Descriptive Analysis of Survey Responses', *Computers and Electronics in Agriculture*, 77: 7-20.

Lesslie, R., Mewett, J. and Walcott, J., 2011, *Landscapes in Transition: Tracking Land Use Change in Australia*, Canberra: DAFF and ABARE.

Midgley, G., 2000, *Systemic Intervention: Philosophy, Methodology and Practice*, New York: Kluwer Academic.

Mullen, J., 2007, 'Productivity Growth and the Returns from Public Investment in R&D in Australian Broadacre Agriculture', *Australian Journal of Agricultural and Resource Economics*, 51(4): 359-84.

Nossal, K. and Lim, K., 2011, 'Innovation and Productivity in the Australian Grains Industry', Canberra: ABARE, *ABARE Research Report 11.06.*

Oreszczyn, S., Lane, A. and Carr, S., 2010, 'The Role of Networks of Practice and Webs of Influencers on Farmers' Engagement With and Learning About Agricultural Innovations', *Journal of Rural Studies*, 26: 404-17.

Pfiefer, C., Jongeneel, R.A., Sonneveld, M.P.W. and Stoorvogel, J.J., 2009, 'Landscape Properties as Drivers for Farm Diversification: A Dutch Case Study', *Land Use Policy*, 26: 1106-15.

Pritchard, B., Burch, D. and Lawrence, G., 2007, 'Neither "Family" nor "Corporate" Farming: Australian Tomato Growers as Farm Family Entrepreneurs', *Journal of Rural Studies*, 23: 75-87.

Richards, S.T. and Bulkley, S.L., 2007, 'Agricultural Entrepreneurs: The First and the Forgotten?', *Center for Employment Policy Entrepreneur Series*, Washington, DC: The Hudson Institute.

Rudwick, V. and Turnbull, D., 1993, 'The Australian Sheep Flock: 1992 Demographics', Canberra: ABARE, *Research Report 93.3.*

Schultz, T.W., 1961, 'Investment in Human Capital', *The American Economic Review*, 51: 1-17.

Tonts, M., 2005, 'Competitive Sport and Social Capital in Rural Australia', *Journal of Rural Studies*, 21: 137-149.

UNCTAD, 2010, 'Agriculture at the Crossroads: Guaranteeing Food Security in a Changing Global Climate', Geneva: United Nations Conference on Trade and Development (UNCTAD). *Policy Briefs 18.*

Villano, R., Fleming, E. and Fleming, P., 2010, 'Evidence of Farm-level Synergies in Mixed-farming Systems in the Australian Wheat-Sheep Zone', *Agricultural Systems*, 103: 146-152.

Wolf, S.A., 2008, 'Professionalization of Agriculture and Distributed Innovation for Multifunctional Landscapes and Territorial Development', *Agriculture and Human Values*, 25: 203-207.

Zhao, S., Sheng, Y. and Kee, H.J., 2009, *Determinants of Total Factor Productivity in the Australian Grains Industry*, Canberra: ABARE.S

Chapter 9

Water Reform in the 21st Century: The Changed Status of Australian Agriculture

Erin F. Smith and Bill Pritchard

Introduction

How is water reform shaping – and re-shaping – rural and regional Australia? What is the importance of water reform in agricultural restructuring? What values should be considered when determining how Australia's water resources are allocated? These questions are not easily answered in contemporary Australia. In the nation's political and social spheres, a set of diverse values, other than agricultural production, contribute toward strategies for managing water resources such as other commercial uses (e.g., mining), the needs of environmental assets, recreational interests and domestic water use. The highly contested nature of these issues in Australia was demonstrated by events during October 2010, when a draft water management plan for the Murray-Darling Basin (the Basin Plan) was made publically available. Representatives of the Murray-Darling Basin Authority, an entity with the mandate for overseeing water resource planning in the Murray-Darling Basin (see Figure 9.1), including the drafting of the Basin Plan, met significant opposition from people in regional and rural areas, culminating in copies of the proposed plan being stacked into bonfires and burnt outside public meetings. Much of this opposition focused on the Plan's proposed reductions, as outlined in the Basin Plan, in the quantity of water available for agricultural use.

Although agriculture is the single biggest user of consumptive water in Australia, 54 percent of total water consumption in 2010-11 (Australian Bureau of Statistics 2012a), in the 21st century it is seen by key policy-makers, and certainly many in the community, as being just one of many different kinds of water uses, that should not be privileged above others. This view represents a significant transformation in the perspectives towards agriculture and water which have historically dominated the Australian polity. Historically, the agricultural industry attracted significant government investment and public subsidies to support its development and expansion. In contemporary Australia, however, the place of water and how it contributes to rural places and people's livelihoods is being re-negotiated with corresponding impacts for the ways in which farmers are able to access and use water.

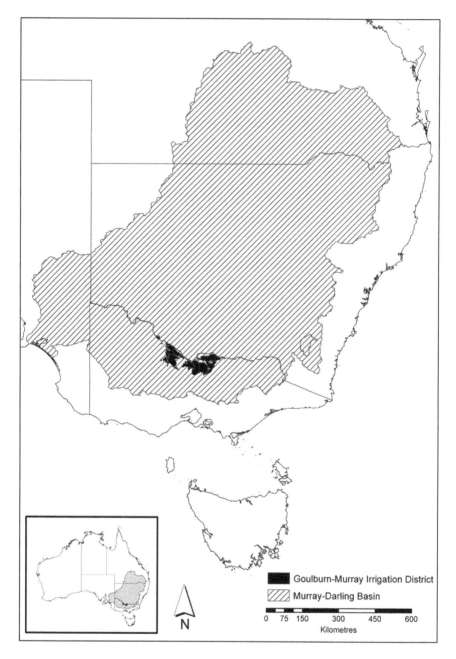

**Figure 9.1   Murray-Darling Basin and Goulburn-Murray
                Irrigation District**

The specific implications of changed water management structures for farming are complex and vary across regions and industries, but it is not too extreme to suggest that the changes represent a profound shift for this industry. The commercial, agricultural underpinnings of farms still remain a dominant and visible social institution in the countryside, notwithstanding much-discussed rural transformations towards post-productivism and multifunctionality (Holmes 2006). While water reform is the focus of this chapter, it is not the only driver of agricultural and rural change. Water reform occurs alongside a complex web of regional and industry restructuring mediated by other economic, social, environmental and political factors, many discussed elsewhere in this book. Recognising the context of these wider policy decisions, then, in this chapter, we trace the Australian water management reforms that have taken place over the last 50 years, noting how they reflect the changed status of agriculture in Australia. Although these reforms are directed at the water industry as a whole, to date they have focused on the Murray-Darling Basin (see Figure 9.1): the largest and most agriculturally important catchment in the country (producing more than one-third of Australia's food supply). Consequently, our discussion also focuses on the Murray-Darling Basin. While all water systems differ in terms of their geography and hydrology, along with the local economic and social milieu which they support, the approach to managing water in the Murray-Darling Basin reflects Australia's approach to water management more generally.

First, we consider the historical importance of water access and irrigation development to the establishment of Australian agriculture, and we highlight the increased importance that water management has within Australian society and politics. Second, we outline the key governance arrangements concerning water ownership and management in Australia, with a particular focus on the Victorian context, recognising the significant change brought about by the entrance of the Commonwealth Government into water management. These changes have legitimised uses of water other than agriculture, and enabled the redistribution of water resources to these other uses, primarily the environment. To conclude the chapter, we provide a brief case study of how these macro-political changes manifest at the local scale.

Context: The Development of Australian Agriculture and the Role of Water

Water resources and water access have been critical for Australian agriculture since European settlement. The development of 'permanent' agriculture, initiated through land selection policies enacted in the final decades of the 19th Century, established a growing population in rural areas. Larger rural populations and the impact of the drought that began in the late 1870s served as impetuses to develop water supply infrastructure and for governing authorities to reconsider water rights. At this time, water rights were founded upon riparian doctrine: the principle that owners of land had a right to make 'reasonable use' of water sources on or adjoining their lands. Large scale irrigation schemes and water storages were progressively developed throughout country Australia, the first at Renmark, South Australia in 1887. These developments established the trajectory of Australian water management over the next 100 years. As argued by Gibbs (2009: 2964), these agendas were guided by an implicit political project characterised by 'colonial patterns of ordering and transforming nature'.

Throughout this period, country Australia held a privileged position within Australia's political and societal sphere; contributing toward the nation's economic prosperity and cultural distinctiveness (Brett 2007). Such a contribution was used to justify substantial government support for the industry through infrastructure development, input subsidies and tariff protection, policies which were maintained for much of the 20th century. Taken together, these constructed what Aitkin (1985) labelled 'country mindedness': a bipartisan political convention to support farming as the economic and cultural bedrock of nation-building. This position of privilege, however, began to be questioned throughout the 1970s and 1980s. Since that time, direct government protection and support for agriculture has gradually been removed, and new ways of managing environmental resources, particularly water, have emerged which challenge the colonial foundations (Gibbs 2009) and agriculture's assumed right to use these resources for production purposes.

Context: The Increased Importance of Water Management in Australia

New approaches to water management were not only driven by changed perspectives toward particular industries. Broader environmental management issues such as addressing the causes and impacts of climate change also gained prominence throughout the final decades of the 20th Century. Thus, efforts to manage water resources more effectively also reflect ways in which countries are attempting to address climate change.

In Australia, the increased political and social importance of water is reflected in changes to Australian federal government ministerial portfolios. In the 1970s, the Australian Government appointed its first Minister responsible for the environment. In the 1980s and 1990s, the importance of this portfolio grew in line with the increasingly complex navigation of national environmental decisions, notably the Tasmania Dams case, the safeguarding of the Queensland Wet Tropics, the mining of uranium at Kakadu National Park, national forest planning, and the management of Antarctica and the Southern Ocean. Then, in 2007, the same year Prime Minister Howard enacted the *Water Act 2007 (Commonwealth)*, Malcolm Turnbull was assigned the Environment and Water Resources portfolio, the first ministerial title to directly refer to water. The title reverted to a more generic environment portfolio under the Rudd-led Labour Government, but water was reinstated in a ministerial title under the Gillard Labour Government in 2010 with Tony Burke being appointed Minister for Sustainability, Environment, Water, Population and Communities. Although seemingly minor, such changes demonstrate the political priorities of successive governments and the increased contemporary significance of water management within Australia. It also reveals the marked change in Australian water governance due to the increased participation of the Federal government, taking over responsibilities that traditionally remained solely with the States. It perhaps even signals the heightened degree of involvement the Commonwealth anticipates having into the future. As will be discussed below, these changes in governance arrangements for water and the environment have brought a return of direct intervention on the part of the federal government in terms of agricultural production.

What must be recognised at the outset, though, is the path dependent nature of approaches to water management, particularly with regard to the Australian context. Australia now has the world's most advanced water market, which is seen as an exemplar for policies elsewhere (Garrick *et al.* 2011). Yet its inception in 1983 as a means for managing water resources seems to have occurred in the absence of public debate as to how water resources are best allocated. In subsequent years, a series of incremental regulatory decisions and the piecemeal development of supporting institutions have reinforced this mode of regulation. This has particular importance to the evolved terms of debate on these issues in the 21st Century. What it has meant is that the scope for offering solutions to the now recognised over-allocation of consumptive water is circumscribed – indeed, delimited – to the requirement of being consistent with the concept of a market. Hence, the focus of debate is no longer upon the overall mechanism of water allocation, but rather is directed toward how best to manage this market to achieve economic, social and environmental outcomes.

Recognition of this path dependency and the limits now imposed on future water management decisions raises questions regarding the relevancy of the Australian approach to other nation states and water resources. For example, to what extent can the 'Australian model' be applied in contexts where rainfall is more consistent than Australia, or where societal attitudes are resistant to the concept of market regulation? While contextual differences will remain, there are common characteristics of water management that need to be addressed in all water management situations: How should limited water resources be allocated to consumptive and environmental uses? How should water ownership be determined? What role do different levels of government have in managing water resources? Each of these questions will be discussed below from the Australian perspective.

Ownership of Australia's Water Resources: Constitutional Arrangements

Water, as a resource, is of great social, economic and environmental importance. Hence, under common law it is common property, 'not especially amenable to private ownership, and best vested in a sovereign state' ("ICM Agriculture Pty Ltd v The Commonwealth" 2009). In Australia, a federation of partially self-governing states, management responsibilities for water courses and water resources reside with the state and territory governments, under powers codified by the Australian Constitution. Negotiations prior to Federation were marked by debates concerning the control of water courses, the outcome of which was crucial for an agreement being reached between the colonies. The outcome of agreement resulted in Section 100 being included in the Australian Constitution in 1900, effectively restricting the powers of the Commonwealth concerning water use and delegating management responsibility to the state and territory governments: 'The Commonwealth shall not, by any law or regulation of trade or commerce, abridge the right of a State or of the residents therein to the reasonable use of the waters of rivers for conservation or irrigation'. These arrangements allowed state governments to implement their own systems of governance for the extraction and use of water passing through their own jurisdictional boundaries, resulting in a series of fragmented water management regulatory systems and practices. What this means for rural Australia, is that the nature of agricultural production systems was not only influenced by place-based characteristics, but was also influenced by the local regulatory regime. In the case of the drier inland areas, water policy and regulation were of great importance for concurrent settlement and agricultural expansion.

Despite these jurisdictional differences, the development of rural Australia was fuelled by a common ideology that viewed agricultural production and intensification as the means by which Australia would achieve national wealth and prosperity. For irrigation regions, then, throughout much of the 20th Century, state and territory governments readily approved licences to extract water, which increased the volume of water diverted from waterways (and ground water sources) and constructed water storages and irrigation infrastructure to facilitate (and encourage) agricultural water use. Thus, irrigators were able to use water largely unrestrained. This provided a basis for agricultural production and its supporting industries tended to dominate rural and regional Australia. It was not until limits to water were imposed, first through government regulation (the Murray-Darling Cap in 1997, see next section) and then through subsequent climatic events (the severest drought for 100 years – approximately 2000-2010) that the legal status of previously issued water licences was tested by the judicial system and the ownership implications of the constitutional arrangements were clarified.

The precise legal nature of water licences is of prime concern for Australian irrigators who design their farming systems and modify their productive practices in response to a myriad factors, including climatic changes, commodity markets, land values, local government regulations, personal and family circumstances and water access. Uncertainty about the reliability of water access (either through natural decrease in rainfall or government intervention) impacts upon the ways in which farmers plan for the future. In turn, this affects the viability of farm systems developed on the basis of sufficient, location-based water availability. Despite there being few instances of case law specifically involving water licenses, those which have taken place reveal the impact of the constitutional arrangements outlined above. One major judicial decision was that of the High Court in *ICM Agriculture v Commonwealth* in which ICM Agriculture, a large Australian owned agricultural company, challenged a decision of the New South Wales (NSW) Government to change their groundwater bore licences originally granted under the *Water Act 1912 (NSW)* with new licences created under the *Water Act 2000 (NSW)*. At stake for the company was a decision that effectively reduced the amount of water they were permitted to use by up to 70 percent. The High Court ruled against the argument that this reduction amounted to an acquisition of property, determining that the NSW Government had not contravened its constitutional responsibilities, and that they had simply modified a statutory right ("ICM Agriculture Pty Ltd v The Commonwealth" 2009). This decision has particular importance to licence holders as it seems to allow governments the scope to reduce water access entitlements without paying 'just terms' compensation (Kildea and Williams 2010). To date, there is limited research that explores farmers' responses to these court decisions and their perceptions with regard to the security of their water entitlements and water ownership more generally. (The term 'water entitlement' is current terminology, replacing previous terminology of water licenses and water rights. A water entitlement is an ongoing entitlement to access to a share of water from a specified consumptive pool.) Nevertheless, it is these constitutional arrangements that frame water management approaches and, as will be discussed in the next section, create challenges when managing inter-jurisdictional water resources.

Management of Australia's Water Resources: Inter-jurisdictional Challenges

The effect of constitutional arrangements is of further contemporary relevance given the difficulties associated with managing inter-jurisdictional river systems. In the Murray-Darling Basin, where the river system passes through four jurisdictions, state governments have long acted in their own interests when regulating the water resources passing through politically constructed boundaries. The Commonwealth was able to manage water courses for the purposes of transportation, trade and commerce, but were specifically not permitted to interfere with irrigation development and, therefore, agricultural expansion pursued by the state governments (see Section 100 of the Australian Constitution quoted above). This feature of past water governance has contributed to a particular kind of 'competitive federalism' whereby individual states seek to optimise their water extractions with little regard for others, leading to tendencies for over-extraction and resultant widespread environmental degradation. The scale of these problems first began to be recognised during the 1980s. Yet despite co-operative efforts in the latter years of the 20th Century, effective reform was slow. The lingering nature of these problems was further exacerbated by the 'Millennium Drought' experienced by much of rural Australia from approximately 2000-2010, although the specific years recognised as drought differ between regions. The drought initiated a degree of water scarcity not experienced by the current generation of farmers, and ushered in heightened complexity with regard to water governance arrangements. These environmental stressors, coupled with political pressure, provided impetus for the Commonwealth Government to directly legislate for water management through the *Water Act 2007 (Commonwealth)*. This legislation over-rode previous strategies based more narrowly on financial incentives and collaborative agreements between the states, as was the case under the National Competition Policy reforms of 1994 (Kildea and Williams 2010). The events of 2007, saw, for the first time, the Commonwealth assert its legislative powers in the management of water. Its legislative intervention in the water management sphere, however, was on the basis of constitutional powers such as trade and commerce, corporations, external affairs, and powers relating to meteorological observations, statistics, weights and measures, rather than state governments referring their water management powers to the Commonwealth. This was insufficient for the Commonwealth to exercise 'its full legal capacity to define the environmental and economic limits for the management of Australia's most significant water resource' (Gardner *et al.* 2009: 105). Thus, additional co-operative agreements were required. In the following year, the *Memorandum of understanding on Murray-Darling Basin Reform* and the *Intergovernmental Agreement on Murray-Darling Basin Reform* were established prior to the passing of the *Water Amendment Act 2008 (Commonwealth)* in which Basin States (those Australian states through which the Murray-Darling river system passes: Queensland, New South Wales, Victoria and South Australia; see Figure 9.1) referred certain powers to the Commonwealth. Thus, in contradiction to narratives which emphasise a gradual 'rolling back' of government involvement in agricultural industries, water reform exemplifies one sphere in which the state has sought to further its reach, albeit within the neo-liberalised contexts of using market mechanisms for managing water resources.

    In practice, then, current Australian water management comprises a complex and dynamic network of relations between the Commonwealth, State and Local Governments. These governance structures have shaped and constrained (and will continue to do so) the nature of water management practices (Kildea and Williams 2010). Today, the terms of production for irrigated agriculture are influenced by intervention on the part of the Commonwealth Government within a regulatory system characterised by the legacy of past water entitlement allotments, attempts to standardise the system across jurisdictions, and the overarching premise of market-based regulation. This system contrasts significantly with the previous state-based system in which farmers' access to water resources was relatively uncontested. The overall impact of these governance arrangements upon individual water users is significant, largely through an administrative system of water licences, water shares and water allocations.

Management of Australia's Water Resources: Administrative Structures

The establishment of water governance arrangements at a national scale is complemented by an equally complex set of administrative structures and legal rights which relate to the ways in which agricultural operators are granted access to water resources. Historically, these structures have differed between jurisdictions (as discussed above), but increased levels of co-operation since 1994 have been directed at creating a strategic framework that is consistent across jurisdictional boundaries. Despite significant moves to standardise a system of water licences, water entitlements and water allocations, there is still more heterogeneity than can be adequately dealt with here, thus the arrangements discussed here are those that refer to the state of Victoria. Nevertheless, the point is that the changes made by state governments were initiated by Commonwealth Government incentives, resulting in spatial changes over where water resources were applied to land.

Water entitlements associated with a given parcel of land were calculated on the basis of land ownership. As irrigation systems were developed and upgraded, further water entitlements were distributed accordingly. When water trading was introduced in 1989 (temporary trade) and 1991 (permanent trade) (Water Act (Victoria) 1989) the logic of water distribution by way of formulae was disrupted. Under the new arrangements of the legislation, land owners' water rights could be bought and sold, but had to be associated, or tied, to a land parcel somewhere. From a geographical perspective, this meant that spatial unevenness was able to develop as the ownership and application of water for irrigation became detached from the strict equal-access metrics of the original area-based calculations.

Water entitlements attached to a land parcel gave rise to water allocations (expressed as a percentage of the water entitlement depending upon the amount of water in storage), which was the physical volume of water that an irrigator could use in a given season. Given the abundance of water resources at the time, allocations were maintained at 100 percent. However, many farmers were able to access additional quantities of water through what was known as 'sales water', which was an additional quantity of water that was allocated after sufficient water was available to meet water needs in the following season. The outcome of this process was that many farm systems were designed to use more water than the entitlement attached to the land, systems which came under severe pressure during the drought period when the quantity of water available for use was limited.

As a further step in the reform process, consistent with the agenda agreed between the Basin States and the Commonwealth Government, in 2007 Victoria allowed for the legal separation of water title from land title. At this time, the previous entitlement system of water rights and sales water was replaced with *water shares* that were designated as either 'high reliability' or 'low reliability'. Even cursory consideration of the change in nomenclature indicates a fundamental shift in what statutory rights for water mean. The earlier terminology referred to water as a 'right', the current terminology refers to one's 'share' of a water resource pool. However, as a statutory right, neither system sanctioned 'ownership' of water as a physical resource; rather, users owned the right to access up to that quantum of water subject to the annual allocation decisions of governing authorities. In a context that now recognised new, legitimate water uses such as the environment, this market system comprising water shares and water allocations influences the way in which limited water resources could be redistributed among these different uses, of which, agriculture is but one use.

Agriculture: One of Many Uses for Australia's Water Resources

Agriculture is no longer a privileged user of Australia's water resources. This represents a fundamental shift in the place that agriculture holds within Australian society and the economy. It also demonstrates broader demographic and economic changes that have taken place in rural and regional Australia as the countryside becomes not only a place of agricultural production, but a place in which other extractive industries are growing (e.g. mining), a place that can – and perhaps should – be consumed and protected, and a place of cultural importance. This changed status was confirmed in the new water policy paradigm introduced during the 1990s. The new policy framework represented a reversal of past approaches to water management which had typically focused upon managing water *supply*; now, the focus was upon managing *demand* (Bjornlund and Rossini 2010). Water users, other than agricultural irrigators, became legitimate users of scarce water resources. The net result was less water being available for agricultural operators which demanded greater water efficiency on farms and created opportunities for farmers to manage their land/water assets and farm systems in new ways. In particular, 'the environment' gained prominence as a worthy recipient of water. The legitimacy of specifically allocating water to the environment was acknowledged by the Howard Coalition Government and played a strong part in Murray-Darling water management by the Rudd and Gillard Labour Governments (2007 onwards). In this section, we review the policy changes that demonstrate this re-organisation of priorities in water management.

Formal recognition of the environment as a legitimate 'user' and recipient of water was made at the Council of Australian Governments (CoAG) meeting in 1994 where state governments agreed to reforms for the water industry, including reforms to resolve river catchments that had been over allocated. The new process would specifically address water pricing, elimination of cross subsidies and transparency in subsidies, allocation of water for the environment, water allocation and entitlement more generally (Council of Australian Governments 1994). Direct action to address the over-allocation of water resources in the Murray-Darling Basin was first imposed the following year, in 1995 when the Murray-Darling Basin Ministerial Council introduced a moratorium on future water extractions. This was followed by 'The Cap' which limited water extractions from many river systems to 1993/94 levels of development (Murray-Darling Basin Commission 1998). Water resources for consumptive uses were now officially limited, heralding the creation of imposed scarcity on this natural resource and ending the era with which agriculturalists could use water without restraint. Yet, basic recognition of legitimacy to use water did little to fundamentally re-distribute water; direct intervention and market participation on the part of the Commonwealth Government was required.

The importance of deliberate management of water resources to meet environmental needs was behind the Commonwealth's intervention into water matters through the *Water Act 2007 (Commonwealth)*. Aside from the marked change that this represented in terms of the Commonwealth's involvement in water management, this legislative move was also significant for the importance placed on effectively managing water for the environment, alongside economic and social needs (Water Act 2007 (Commonwealth)). However, the maximisation of 'the net economic returns to the Australian community from the use and management of the Basin water resources' (Water Act 2007 (Commonwealth) Section 3d(iii)) is subject to ensuring that unsustainable levels of water extraction are returned to environmentally sustainable levels and, also, contingent upon the protection, restoration and provision of ecological values and ecosystem services. In sum then, the changed legislative landscape prescribed a situation whereby people living in rural areas, especially those living in areas highly dependent on water resources, would have access to fewer water resources than they did in the past.

To enact this legislation, the Water Act also provided for the formation of the politically independent Murray-Darling Basin Authority (MDBA), subsuming the responsibilities of the former Murray-Darling Basin Commission. Most crucially – at least in the short term – the MDBA was charged with preparing a strategic plan for the management of water resources in the Basin that would convert the aspirations of legislation into concrete decisions. This involved the development of defensible long-term average sustainable diversion limits for the Basin and its parts, limits which 'must reflect an environmentally sustainable level of take' (Water Act 2007 (Commonwealth) Section 21). Within these terms of reference the MDBA prepared *The Guide to the Basin Plan* in which the Authority proposed to reduce annual consumptive water use by 3,000 to 4,000 gigalitres (1 gigalitre (GL) = 1 billion litres), out of a total average level of 15,400 GL across the whole Basin (Murray-Darling Basin Authority 2010a). This proposal enunciated a clear break from the existing water governance regime, established through 'The Cap' in 1997. Whereas the purpose of 'The Cap' was to prevent the quantity of water extracted from increasing (Murray-Darling Basin Commission 1998), the 2010 proposals specified a deliberate decrease in water diversions.

Authority representatives introduced their proposals to rural communities in October 2010 through a series of public meetings at which irrigators and others living in these areas voiced their intense opposition to the proposals. They contended that the economic and social impacts of the proposal had not been fully examined. Clearly, much was at stake. The strength of the response from rural Australia resulted in the Commonwealth ordering a revision of the plan and the Commonwealth House of Representatives forming of a Standing Committee

Inquiry ('The Windsor Inquiry') into the impact upon regional Australia should the proposed Basin Plan be implemented (House of Representatives Standing Committee on Regional Australia 2011). The revised plan was signed into law by the Minister for Sustainability, Environment, Water, Population and Communities, Tony Burke, in November 2012.

Preceding the release of the final Basin Plan and empowered by the *Water Amendment Act 2008 (Commonwealth)*, the Commonwealth embarked on a large-scale plan of water buybacks, via a tender process, from willing sellers (largely agricultural irrigators). The programme – titled *Restoring the Balance* – is part of the Commonwealth's long-term *Water for the Future* initiative to 'better balance the water needs of communities, farmers and the environment' (Department of Sustainability, Environment, Water, Population and Communities 2010a). As at 30 November 2012, 1,331 GL of entitlements had been secured, including an annual average volume of 1,094 GL available for the environment (Department of Sustainability, Environment Water, Population and Communities 2012). For rural Australia, the delivery of the Basin Plan and commencement of water buybacks represented direct government intervention into the market-based terms of production following the gradual withdrawal of government support and subsidies for agricultural production under which irrigators have been operating since the 1980s. Indeed, the Commonwealth Government views the *Water for the Future* programme as 'the largest single agricultural adjustment program in Australia's history' (Australian Government 2011: 2).

The involvement of the Commonwealth Government in agricultural restructuring via water markets is demonstrated through a key institutional change that was necessary to enable *Restoring the Balance* to be implemented. The Commonwealth Environmental Water Holder (CEWH) was established to manage the Commonwealth's water resources for protecting or restoring 'the environmental assets of the Murray-Darling Basin and other areas outside the Murray-Darling Basin where the Commonwealth holds water' (Water Act 2007 (Commonwealth) Sections 104 and 105). To achieve this, the CEWH is able to exercise 'any powers of the Commonwealth to purchase, dispose of and otherwise deal in water and water access rights, water delivery rights or irrigation rights' (Water Act 2007 (Commonwealth) Section 105). The creation of the CEWH effectively introduces the Commonwealth Government, on behalf of the environment, as an active participant in the water market, unlike the previous situation in which the Government's role was primarily regulatory.

The strategy of reducing diversions through buybacks is augmented by water efficiency infrastructure spending from the Commonwealth. The co-existence of these two arms of water governance represents quite differing – some may say contradictory – approaches to sustainability within the Basin. Water efficiency spending is undertaken through the *Sustainable Rural Water Use and Infrastructure Program* (SRWUIP), which was budgeted $5.8 billion with an emphasis on renewal of irrigation infrastructure on and off-farm (Department of Sustainability, Environment, Water, Population and Communities 2010b). So at the same time that the Commonwealth (through the CEWH) is purchasing water entitlements and, thus, reducing the quantum of water potentially diverted from the river courses of the Basin, it is expending resources on those who retain their entitlements so that they can use water more efficiently.

Case Study: Goulburn-Murray Irrigation District, Victoria

Prior to concluding this chapter, a brief case study of one irrigation area – Goulburn-Murray Irrigation District, Victoria – is presented to illustrate the ways in which the macro-political changes outlined above manifest at the local scale, most notably through the water buyback programme which provided opportunities for farmers to sell all or part of their water shares at a time when agriculture was under severe pressure from the drought, enabling them to remain farming or exit farming altogether (Cheesman and Wheeler 2012).

The Goulburn-Murray Irrigation District (GMID) is located in northern Victoria. Irrigation water and infrastructure is managed by Goulburn-Murray Water (GMW), a statutory corporation which manages irrigation systems, primarily flowing through two major water systems: the Goulburn System, sourced from Lake Eildon; and the Murray System, sourced from Lake Hume and the Dartmouth Dam. Agriculture in the region is dominated by dairying and horticulture. Officially the drought was considered to have begun in this region in 1997 (Department of Sustainability and Environment 2009). Since then, this region has undergone significant change in terms of the structure of the agricultural industry and land use, influenced by the changed priorities for water and governance arrangements outlined above. Water reform alone has not created these changes in a causal sense; indeed, it is impossible to disentangle the exact impacts of particular policy shifts from broader contextual factors, in particular the drought and commodity markets. What seems apparent though is that the heightened degree of Commonwealth intervention in water allocation has played a role in altering the nature of this region, and provided irrigators with opportunities to modify their water ownership.

The effects of the drought were not immediately felt because of the large-scale water storages which delayed reductions in water allocations until 2002 for farmers connected to the Goulburn System and 2006 for farmers on the Murray System (Goulburn-Murray Water 2012a). It would be another four years, however, before significant rainfall and storages were replenished. In the intervening years the Commonwealth Government commenced its water buyback programme, issuing the first tender in which GMID irrigators could participate in February 2008.

Water market activity, including government purchases of water for the environment and private sales, alongside water savings achieved through irrigation modernisation, have resulted in a net reduction of water shares held within the GMID. Prior to permanent water trading, there were 1,620 GL of high reliability water shares in the GMID; this had decreased slightly by 1998 (just prior to the drought) to 1,601 GL. Fourteen years later (April 2012), 1,096 GL of high reliability water shares remained, representing a 35 percent reduction since immediately prior to the drought (Goulburn-Murray Water 2012b). The Commonwealth Environmental Water Holder has secured approximately 60 percent (299.5 GL as at January 2012) of the water that has been traded out of the GMID (Goulburn-Murray Water 2012b). Although this process was voluntary on the part of irrigators, it demonstrates a marked shift in the way in which the Commonwealth Government interacts with irrigators, from a position where it encourages water use and agricultural production, to one where it positions itself as a purchaser of those same irrigator water entitlements that were originally distributed by government to encourage agricultural production. The net effect is to reduce the consumptive pool of water available to irrigators, although additional factors such as agricultural industry, broader local economies and climatic changes will influence which regions experience a net increase or net decrease in the amount of water held by irrigators.

The combination of drought and reduced consumptive water (combined with fluctuations in commodity markets) has resulted in 'significant structural change in the location and type of irrigated agriculture throughout the GMID' (HMC Property Group 2010: 5). Dairying and horticultural enterprises have been the dominant water users throughout the region during the drought, while mixed farming operations have declined significantly (Murray-Darling Basin Authority 2010b). Furthermore, a study which investigated changing land use in the GMID between 2006 and 2010 revealed that the most marked change was the amount of land now lying idle, that was previously operated for dairying purposes. The reduction in working dairies across the region is estimated to have been approximately 29 percent across the four year timeframe under investigation (HMC Property Group 2010). Changes to dairy farm systems have also occurred, such as an increased reliance on annual crops and pastures, rather than the traditional system of permanent pasture. Such changes alter the volume and timing of farmers' water requirements (Murray-Darling Basin Authority 2010b).

In addition, the landscape is being fundamentally reconfigured via a Commonwealth and Victorian Government funded irrigation modernisation programme in order to reduce the irrigation footprint through decommissioning under-utilised assets and upgrading to more efficient water delivery infrastructure (Northern Victoria Irrigation Renewal Project 2012). This irrigation modernisation project was an outcome of the Murray-Darling Reform Intergovernmental Agreement in 2008. In the second phase of this project, the Commonwealth agreed to contribute funding on the basis of being able to purchase for the environment the total anticipated water savings (approximately 204 GL) The 50 percent reduction of the length of irrigation channelling achieved through this project contrasts starkly to the large-scale government-sponsored expansion of agricultural which established the infrastructure in this region throughout the first half of the 20th Century. Reducing the irrigation footprint and re-distributing water to the environment demonstrates the ideological shifts toward water management and distribution that have occurred in the last ten years and which will determine the possible paths for agricultural restructuring in the future.

Conclusion

Australian agriculture has been subject to myriad pressures for some time – key among them has been access to water. Traditional post-war challenges such as escalating costs of production seem likely to continue for farm operators, but new challenges are also surfacing such as the much more recently contested nature of natural resources as other, non-productive values, such as the environment, gain increased legitimacy and shape policy for rural and regional Australia. The societal changes and the ensuing regulatory modifications associated with water management that have been discussed in this chapter occurred at the same time as agriculturalists experienced a severe, prolonged period of drought. Ascribing causality between these events is unhelpful, nevertheless the limits to water resources imposed by climatic conditions no doubt expedited the actions of the Commonwealth Government, creating a much more complex governance structure in which agriculturalists needed to operate.

Once considered the foundation on which national prosperity was built and a significant contributor to the national identity, the role of agricultural production in Australia has changed and it will likely undergo further change in the future. Agriculture has been re-positioned as one of many users of environmental resources. Yet, just over half of Australia's land area is used for some form of agricultural production (Australian Bureau of Statistics 2012b), so its role and importance in resource allocation debates remain. The environmental agenda is being worked and re-worked through shifting societal values and changes to governance structures. Indeed, it should be noted that comprehensive resolutions have not been met in all rural areas and for all agricultural operators. Rather, these reforms are ongoing, and many more significant decisions are yet to be made.

The Australian experience to date, demonstrates the path-dependent and constrained nature in which contemporary water reform can take place. Practical advancement of the environmental agenda to address historical over-allocation of water resources required government intervention at the Federal level. Thus, the neoliberal economic agenda pursued by the Australian Government since the 1980s, has been applied to the management of the nation's resources. Legislation has empowered the Commonwealth Government as a market participant to re-distribute water from agriculturalists to the environment. Somewhat differently, substantial government support by way of subsidised infrastructure programmes has re-emerged in the context of new ways to manage water (and many other environmental resources) which are challenging the long assumed right to their use on the part of agriculture. 'Patterns of ordering and transforming nature' remain, but within a different paradigm, in which the environment is prioritised. These events of the first decade in the 21st Century coalesced to create vastly different terms of production for agriculturalists. Indeed, the extent of regulatory change and the on-going re-organisation of water governance structures have potentially created a vastly different rural and regional Australia for the future.

References

Aitkin, D., 1985, '"Countrymindedness" – The Spread of an Idea', *Australian Cultural History*, 4: 34-40.

Australian Bureau of Statistics, 2012a, *Water Account Australia: 2010-11*. Canberra: Australian Bureau of Statistics. [Online]. Available at: http://www.abs.gov.au/AUSSTATS/abs@.nsf/Lookup/4610.0Main+Features302010-11?OpenDocument [accessed: 08 January 2013].

Australian Bureau of Statistics, 2012b, *Agricultural Commodities Australia: 2010-11*, Canberra: Australian Bureau of Statistics.

Australian Government, 2011, *Australian Government Response to the House of Representatives Standing Committee on Regional Australia Committee Report: Of Drought and Flooding Rains. Inquiry into the Impact of the Guide to the Murray-Darling Basin Plan in Regional Australia*. [Online]. Available at: http://www.environment.gov.au/water/publications/mdb/pubs/windsor-inquiry-response.pdf [accessed: 27 November 2012].

Bjornlund, H. and Rossini, P., 2010, *Climate Change, Water Scarcity and Water Markets: Implications for Farmers' Wealth and Farm Succession*, 16th Pacific Rim Real Estate Society Conference, Wellington, Available at: http://www.prres.net/papers/Bjornlund_Climate_Change_water_scacity_water_markets-Implications_farmers_wealth_farm_succession.pdf [accessed: 08 January 2013].

Brett, J., 2007, 'The Country, the City and the State in the Australian Settlement', *Australian Journal of Political Science*, 42: 1-17.

Cheesman, J. and Wheeler, S., 2012, *Survey of Water Entitlement Sellers Under the Restoring the Balance in the Murray-Darling Basin Program: Final Report Prepared for the Department of Sustainability, Environment, Water, Population and Communities*, Melbourne: Marsden Jacob Associates Pty Ltd.

Council of Australian Governments, 1994, *Communique. Meeting of COAG in Hobart 25 February, 1994: Attachment A – Water Resource Policy* [Online]. Available at: http://www.environment.gov.au/water/publications/action/pubs/policyframework.pdf [accessed: 24 August 2012].

Department of Sustainability and Environment, 2009, *Northern Region Sustainable Water Strategy*, Melbourne: State of Victoria, Department of Sustainability and Environment.

Department of Sustainability, Environment, Water, Population and Communities, 2010a, *Water for the Future: Factsheet*, Canberra: Department of Sustainability,

Environment, Water, Population and Communities. [Online]. Available at: http://www.environment.gov.au/water/publications/action/pubs/water-for-the-future.pdf [accessed: 21 September 2012].

Department of Sustainability, Environment, Water, Population and Communities, 2010b, *Sustainable Rural Water Use and Infrastructure Program: Factsheet*, Canberra: Department of Sustainability, Environment, Water, Population and Communities. [Online]. Available at: http://www.environment.gov.au/water/publications/policy-programs/pubs/srwui-program-factsheet.pdf [accessed: 10 January 2013].

Department of Sustainability, Environment, Water, Population and Communities, 2012, *Progress of Water Recovery Under the Restoring the Balance in the Murray-Darling Basin program* [Online]. Available at: http://www.environment.gov.au/water/policy-programs/entitlement-purchasing/progress.html [accessed: 09 January 2013].

Gardner, A., Bartlett, R., Gray, J., and Carney, G., 2009, 'The Constitutional Framework for Water Resources Management', in A. Gardner, R. Bartlett, J. Gray and J. Carney (eds), *Water Resources Law*, Chatswood: LexisNexis Butterworths, 81-105.

Garrick, D., Lane-Miller, C. and McCoy, A.L., 2011, 'Institutional Innovations to Govern Environmental Water in the Western United States: Lessons from Australia's Murray-Darling Basin', *Economic Papers*, 30(2): 167-84.

Gibbs, L., 2009, 'Just Add Water: Colonisation, Water Governance, and the Australian Inland', *Environment and Planning A*, 41: 2964-2983.

Goulburn-Murray Water, 2012a, *Resources: Historical Allocation Data* [Online]. Available at: http://www.nvrm.net.au/resources.aspx [accessed: 3 September 2012].

Goulburn-Murray Water, 2012b, *Goulburn-Murray Water Proposed Murray-Darling Basin Plan Submission* [Online]. Available at: http://www.g-mwater.com.au/downloads/media-releases/Goulburn-Murray_Water_Proposed_Murray_Darling_Basin_Plan_Submission.pdf [accessed: 20 September 2012].

HMC Property Group, 2010, *Changing Land Use in the GMID 2006-2010: Where Have All the Dairies Gone?* Shepparton: HMC Property Group. [Online]. Available at: http://www.greatershepparton.com.au/downloads/planning/regional_rural_land_use_strategy/reference%20docs/Changing%20Land%20Use%20in%20the%20GMID.pdf [accessed: 27 August 2012].

Holmes, J., 2006, 'Impulses Towards a Multifunctional Transition in Rural Australia: Gaps in the Research Agenda', *Journal of Rural Studies*, 22: 142-160.

House of Representatives Standing Committee on Regional Australia, 2011, *Of Drought and Flooding Rains: Inquiry into the Impact of the Guide to the Murray-Darling Basin Plan*. Canberra: Commonwealth of Australia. [Online]. Available at: http://www.aph.gov.au/Parliamentary_Business/Committees/House_of_Representatives_Committees?url=ra/murraydarling/report.htm [accessed: 21 September 2012].

ICM Agriculture v Commonwealth. (2009) 240 CLR 140[55]. High Court of Australia.

Kildea, P. and Williams, G., 2010, 'The Constitution and the Management of Water in Australia's Rivers', *Sydney Law Review*, 32: 595-616.

Murray-Darling Basin Authority, 2010a, *Guide to the Proposed Basin Plan: Overview (vol. 1)*. Canberra: Murray-Darling Basin Authority. [Online]. Available at: http://download.mdba.gov.au/Guide_to_the_Basin_Plan_Volume_1_web.pdf [accessed: 21 September 2012].

Murray-Darling Basin Authority, 2010b, *The Guide to the Proposed Basin Plan: Appendix C: Goulburn Murray Community Profile*, Canberra: Murray-Darling Basin Authority. [Online]. Available at: http://download.mdba.gov.au/AppendixC_Goulburn_Murray_community_profile.pdf [accessed: 09 January, 2013].

Murray-Darling Basin Commission, 1998, *Murray-Darling Basin Cap on Diversions, Water Year 1997/98: Striking the Balance.* Canberra: Murray-Darling Basin Commission. [Online]. Available at: http://www2.mdbc.gov.au/__data/page/81/Striking_the_Balance_Report_97_98.pdf [accessed: 21 September 2012].

Northern Victoria Irrigation Renewal Project, 2012, *Annual Report: 2011-2012*, Shepparton, Victoria: Goulburn-Murray Rural Water Corporation.

Water Act, 1989, (Victoria). Melbourne, Victorian Legislation and Parliamentary Documents.

Water Act, 2007, (Commonwealth), Act No. 137, Canberra, ComLaw.

Water Amendment Act, 2008, (Commonwealth), Act No. 139, Canberra, ComLaw.

Chapter 10

Rural Place Marketing, Tourism and Creativity: Entering the Post-Productivist Countryside

Chris Gibson

There is more to rural Australia than agriculture. Broadscale transitions in the economy have meant that farming (like manufacturing) has become more capital rather than labour intensive, while education, tourism, finance, and health and community services have become significant employment sectors (Table 10.1). Economic activities centred on leisure, consumption and the creative arts are increasingly important. Even accounting for recent contraction, tourism is Australia's largest export earner, worth around $34 billion annually (Hooper and van Zyl 2011), while the creative industries are worth somewhere between A$20 and A$25 billion per annum, a figure comparable to that of the residential construction and road transport sectors (Gibson *et al.* 2002, Higgs *et al.* 2007). Such developments have stimulated efforts to reorientate regional economic development and employment away from 'traditional' bases towards a mix of leisure, lifestyle and creative activities.

The transition from agriculture to service industries, leisure and consumption has been described as 'post-productivism' – a move away from mass production of agricultural commodities, towards rural lifestyles and landscapes as *the* marketable or defining features (Halfacree 1997, Hopkins 1998, Kneafsey 2001). Agriculture is no longer the backbone of the rural economy, and farming landscapes are instead commodified as 'scenery' for the tourism industry, or for real estate developments. Some 45 percent of total tourism expenditure in Australia is undertaken outside capital cities (Hooper and van Zyl 2011). Rural areas have much to gain from such a transition, though it is only partially a departure from previous eras.

The pressures to embrace change are considerable. Rural regions have had to grapple with substantial shifts in agricultural organisation (see also Chapters 8 and 9). Rural places have been plagued by the intertwining problems of ageing and population stagnation, especially the loss of young people; (see Chapter 2), employment decline, increasing climatic variability and changes in the structure and competitiveness of the global agrifood sector. Mining has provided alternative growth in some regions; although its employment multiplier effects are low, compared with services and tourism. What economic

Table 10.1 Employment Change in Non-Metropolitan Australia,* by Selected Industries, by State

	Agriculture, Forestry and Fishing			Mining			Construction			Retail Trade			Accommodation and Food Services		
	2001	2006	2011	2001	2006	2011	2001	2006	2011	2001	2006	2011	2001	2006	2011
NSW	8.7%	7.0%	5.8%	1.4%	1.7%	2.5%	6.9%	7.8%	7.9%	15.9%	12.3%	11.4%	6.0%	7.5%	7.7%
Victoria	11.2%	9.2%	7.8%	0.4%	0.6%	0.8%	6.7%	8.2%	8.9%	15.3%	12.2%	11.6%	4.9%	6.3%	6.8%
Queensland	8.4%	5.8%	4.6%	1.9%	2.6%	3.9%	7.4%	9.9%	9.7%	15.6%	12.1%	11.3%	6.8%	8.0%	7.9%
South Australia	18.6%	15.2%	14.3%	1.4%	1.7%	3.0%	5.8%	6.8%	7.1%	13.5%	11.2%	10.8%	4.8%	6.1%	6.8%
Western Australia	13.3%	11.0%	8.9%	8.2%	7.7%	11.2%	7.9%	9.6%	10.9%	12.9%	10.5%	9.1%	5.1%	6.0%	6.2%
Tasmania	6.7%	5.6%	7.3%	0.8%	0.8%	2.1%	5.1%	6.7%	7.7%	15.0%	11.9%	11.2%	5.2%	6.9%	7.4%
Northern Territory	4.7%	4.1%	3.6%	4.6%	2.6%	3.7%	5.8%	5.6%	6.5%	9.6%	8.0%	7.4%	7.0%	7.1%	6.5%
ACT	0.4%	0.3%	0.2%	0.0%	0.1%	0.0%	4.8%	5.4%	6.1%	12.2%	8.7%	7.6%	5.0%	5.8%	5.8%
Australia	9.3%	7.3%	6.1%	1.8%	2.1%	3.1%	6.8%	8.3%	8.6%	15.0%	11.8%	10.9%	5.8%	7.1%	7.3%

	Combined selected services**			Public Administration and Safety			Education and Training			Health Care and Social Assistance			Arts and Recreation Services		
	2001	2006	2011	2001	2006	2011	2001	2006	2011	2001	2006	2011	2001	2006	2011
NSW	-	9.1%	9.5%	4.7%	6.9%	7.2%	7.7%	8.3%	8.6%	10.4%	11.5%	13.0%	1.8%	1.2%	1.2%
Victoria	-	8.2%	8.6%	3.3%	5.8%	6.1%	7.5%	7.9%	8.2%	10.8%	11.4%	13.1%	2.0%	1.2%	1.3%
Queensland	-	9.0%	9.8%	4.8%	6.3%	6.0%	7.3%	7.4%	7.7%	9.0%	9.6%	11.4%	2.4%	1.4%	1.5%
South Australia	-	7.4%	6.9%	3.4%	4.9%	4.8%	6.3%	6.9%	7.0%	9.0%	10.4%	11.5%	1.4%	0.7%	0.8%
Western Australia	-	7.0%	7.1%	4.8%	6.0%	5.8%	6.8%	7.3%	7.6%	7.4%	8.5%	8.4%	1.3%	0.7%	0.7%
Tasmania	-	9.6%	8.3%	5.3%	8.4%	6.5%	8.2%	8.3%	8.4%	11.6%	11.4%	11.3%	2.4%	1.4%	1.3%
Northern Territory	-	7.2%	7.1%	20.6%	22.6%	19.9%	8.2%	9.1%	11.0%	9.5%	12.6%	11.4%	2.7%	1.8%	2.8%
ACT	-	14.0%	13.9%	23.9%	30.1%	32.7%	8.9%	9.0%	8.8%	8.3%	8.8%	9.2%	3.6%	1.8%	1.7%
Australia	-	9.0%	9.3%	5.6%	7.8%	8.0%	7.5%	7.9%	8.2%	9.8%	10.5%	11.9%	2.1%	1.2%	1.3%

* Non-metropolitan areas defined as those outside capital cities in each state. Data generated using ABS Statistical Area boundaries (e.g. 'Rest of NSW')
** Combined finance, insurance, professional, scientific, technical, administrative and support services categories. Significant reclassification of service industry categories for the 2006 census prevents meaningful concordance with 2001 data, hence are not included here.

Source: ABS 2001, 2006, 2011

benefits do flow from mining have been geographically uneven (compare for instance, mining employment in rural Western Australia and Victoria, in Table 10.1), and are mitigated by increasing use of fly-in, fly-out labour, which brings with it a host of social tensions (see Chapter 7). Agriculture may well remain a sizeable export industry, and an important historical and cultural feature of rural Australian life, but neither it nor mining can sustain rural employment indefinitely.

Across Australia, rural places have experimented to varying degrees with creative means to generate economic activity and employment. Towns within striking distances of capital cities (Berry and Kiama in NSW, Castlemaine in Victoria, Victor Harbor in South Australia) have presented themselves as weekender destinations or as places to live and telecommute (so long as information technology infrastructure is adequate). With lifestyle 'tree change' migration has come renewed viability, but also problems of housing affordability (Costello 2007). Other more distant settlements have had to work harder to bring in people and revive economies. Construction of 'flagship' tourist attractions has been one avenue. Longreach for instance opened both the Stockman's Hall of Fame (which over a million people have visited since its opening in 1988) and the Qantas Founders Outback Museum (the airline was founded in the region in the early 1920s). Nearby Winton opened a Waltzing Matilda museum (the song was first performed there in 1895). In the shadow of a declining mining industry, Broken Hill promoted itself as a visual arts hub. Young, in south-west NSW, took in Afghan refugees to work in the local meatworks (see Chapter 4). The tiny town of Cumnock (NSW, population 288) pioneered a scheme whereby new families moving to the area could rent farmhouses for $1 per week – a scheme as much about good media coverage as housing affordability. That scheme subsequently spread to fourteen other tiny, struggling towns.

A mix of post-productivist activities – tourism, leisure, adventure and sporting activities, festivals, creative arts, services and cultural industries – are increasingly part of the regeneration strategies of rural places. Marketing has become as important as the production of physical goods, and beyond the 'hard' outcomes of jobs and incomes, intangibles such as place association and emotion are now part of broader regional development equations. Rural images, lifestyles and landscapes become saleable assets. The hope is to use rurality itself to generate new forms of inward investment, attract potential in-migrants, bolster a distinctively rural form of creativity and innovation, and more generally provide a better quality of life. In parallel to declining industrial places struggling with images of rust and decay, in rural regions the tactic is to counter negative representations of decline by building up new positive associations, through directly experiencing that place. That might require attracting tourists in the first instance, and creating a nascent emotional bond with a place that could later be exploited through repeat tourism visits, word-of-mouth effect, and even more permanent migration. This chapter reviews such evolving strategies.

Marketing the Countryside

Rural areas face a public relations conundrum. Although rural Australia has for over a century been considered a heartland of national values, a healthy place associated with myths of mateship, countrymindedness, community and honesty (Botterill 2006), a contemporary struggle is to present a convincing image to the outside world as modern, diverse, cosmopolitan. Although 'country' evokes a whole swag of positive connotations, often nostalgic, associated with a simple life, rusticity and quietude (see Chapter 11), the risk is to come across as parochial or backward (Gibson and Davidson 2004). The stigma of being cast as 'hick' remains ever possible.

Strategies to market rural places to both tourists and potential new migrants have thus diversified (Connell and McManus 2011). Some have tapped into historical portrayals of 'country' hospitality and values, and yet sought ways to appear contemporary. Tamworth (NSW) claimed to be both 'country music capital' (through its famous festival and national awards ceremony) and 'cosmopolitan country' (Gibson and Davidson 2004). Moree (NSW) and Daylesford-Hepburn Springs (Victoria) revived the early twentieth century marketing of their mineral spas, mobilising notions of heritage and rural retreat alongside narratives of sophistication and bodily rejuvenation, targeting urbane baby boomers and gay and lesbian couples (Gorman-Murray *et al.* 2012, White 2012). Both Orange (NSW) and Mildura (Victoria) sought to become 'foodie' destinations with gourmet restaurants and food and wine festivals. The self-proclaimed 'Rural City of Wangaratta' (Victoria) made a conscious decision to host an avant-garde jazz festival, in time becoming known as the nation's most respected event within what is otherwise imagined to be an urbane, metropolitan musical genre (Curtis 2010). Even such tiny towns as Milawa (Victoria, population 543) have got in on the act, marketing itself as the 'hub of the Milawa Gourmet Region', with vineyards, a cheese company, boutique mustards, a regular farm producers market, and a winter wine and food festival (http://www.milawagourmet.com/). So too has tiny, isolated King Island (Khamis 2007). Imagined distinctions between city and country can be dissolved.

Overtures to the rural idyll and bourgeois tastes are more difficult to make elsewhere: in working-class places, abattoir towns, or those that suffer from memories of tragic events. Snowtown, in South Australia, was where the remains of eight bodies were found in 1999 in barrels of acid in a disused bank vault. Forever associated with the murders (although the atrocities did not actually occur in the town, but nearby), and even the subject of a 2011 feature film of the same name, it is hard to imagine a place with worse prospects of attracting tourists or newcomers. Snowtown embodies rural dystopia – the antithesis of harmony, tranquillity and safety (Rofe 2013). It has nevertheless become a cult destination of sorts for adherents of 'dark tourism'. Curious tourists are drawn to the town by its murderous notoriety – leading one local proprietor to begin selling ceramic body-in-a-barrel ornaments and fridge magnets ('Snowtown SA – you'll have a barrel of fun'). In her words,

> It was a terrible thing but we can't do anything about it, so we have to move on with life. You can't pretend it's not there or pretend it didn't happen so I'm doing something that can bring money into Snowtown. Every day cars stop and take a photo of the bank, while some tourists often tell me they would like to see the bank vault and we also get a lot of motorcyclists who park their vehicle in front of the building to have their picture taken. Tourists visiting the town were wanting something more than fridge magnets and spoons (quoted in Kennett 2012)

Moe (Victoria) has similarly become a 'dead zone' after heinous crimes (Hess and Waller 2012). Other places sought safer options. Bowral, already wealthy (but with declining agricultural employment and an ageing population), encouraged upscale clothing boutiques and built a museum in honour of cricket legend Don Bradman; while Parkes built a museum around Elvis Presley (on the back of the town's famous Elvis Festival). Emblematic of the clash of old and new, the Elvis museum – replete with gold lamé suits and other rock 'n' roll memorabilia – was built immediately next door to the older, traditional local museum. Meanwhile Broome (WA) commodified its history as a Japanese pearling centre, in the process seeking to reconcile 'the dark side of the early history of pearling (high death rates, the forced labour of Aborigines, use of Asian indentured workers...)', with 'the glamour and attractiveness of pearls and Broome as a tropical resort town' (Frost 2004: 281). Dark pasts give way to idyllic futures.

Reinventing Rural Tourism

Tourism strategies have similarly evolved. The old 'holidays in the country' are a thing of the past. Australians by and large hug the coast or go overseas. Half the battle can be just to get people to visit in the first place. As with the overall tourism industry, domestic travel to regional areas has stagnated, and the high Australian dollar has kept international arrivals at the lowest levels in a decade (Figure 10.1). Although tourism has become a more prominent part of the employment mix, it has not provided the easy panacea to regional restructuring that some hoped it to be. Seasonality is a basic problem: stiflingly hot inland regions must compete with major domestic coastal destinations in summer, and are too cold in winter (though not cold enough to guarantee snow, and thus skiing and snowboarding possibilities). Even the Snowy Mountains is threatened, both by climate change (hence resorts now artificially produce snow) and by the strong Australian dollar, which lures Australian skiers to the now more affordable, and deeper, snow fields of Japan and New Zealand.

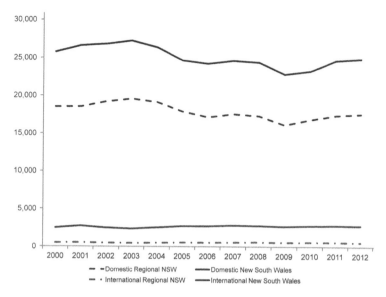

**Figure 10.1   Overnight Domestic and International Visitors,
Regional New South Wales, 2000-2012 ('000)**

Source: Destination NSW (2012)

Rural areas therefore need more interesting possibilities and more targeted means to attract visitors. Not all can succeed. Attempts to develop wild-life tourism in a remote part of Western Australia failed (Hughes and Macbeth 2005) and book towns have had only short-lived success (Kennedy 2011). One distinctively Australian response has been for regional towns to construct 'big things' – giant statues of a local icon or food product, usually adjacent to a major petrol station/ visitor centre/restaurant complex. Early examples – Coffs Harbour's Big Banana (1964), the Sunshine Coast's Big Pineapple (1971), Robertson's Big Potato (early 1970s), Adaminaby's Big Trout (1973), Kingston's Big Lobster (1979) – were built on the back of increasing affluence and the growth in domestic tourism enabled by newly improved interstate highways. They were followed by a slew of others as the lucrative potential of the tourism dollar became more obvious – folk art, heritage and quirky creativity combined.

Some of the more iconic (or simply weird) include Goulburn's Big Merino (Rambo), Tamworth's Big Golden Guitar, Guyra's Big Lamb, Ballina's Big Prawn, Humpty Doo's Big Boxing Crocodile, Mission Beach's Big Cassowary, Tully's Big Gumboot, Bowen's Big Mango, Bundaberg's Big Rum Bottle, Kimba's Big Galah, Churchill's Big Cigar, Swan Hill's Giant Murray Cod, and Gippsland's Giant Earthworm (Clark 2004). Glen Innes (NSW) revived its Scottish heritage and built circles of standing stones, a Gaelic facsimile of sorts (Connell and Rugendyke 2010). Barellan (NSW) erected the 14 metre-long Big Tennis Racquet in honour of Evonne Goolagong, who grew up there. Winton has the world's largest deckchair (in a permanent open air cinema). Evocative of the shifting relationship between agriculture and tourism, Tailem Bend (SA) in 2009 opened The Big Olive at its olive processing plant and visitor centre. Usually the hope is to lure passers-by to stop en route and spend a few dollars locally – although rather more ambitiously, Murweh Shire Council, in remote, inland Queensland, anticipated 'thousands of tourists' to its Big Meat Ant (opened in 2011) (Arthur 2011). In contrast, and with no tourism agenda in mind, local residents in Kiama (NSW) built a Big Poo in 2002 as a protest against Sydney Water's decision not to recycle waste water in the region. Meanwhile Dunedoo (NSW) could not bring itself to open a Big Dunny (toilet); and Albion Park, voted in an online poll as 'Australia's most bogan place', seriously considered, but decided not to go ahead with a Big Ugg Boot (Gibson 2013). Only some icons are palatable.

Specialisation and strategic tourism marketing have become more common. Echuca – the closest point on the Murray River to Melbourne, and hence once Australia's largest inland port – has been a site of heritage tourism for some time, operating the world's largest fleet of operating paddle steamers, evoking the glory days of riverboat tourism. Increasingly, however, it targets Melbournians for weekender visits for floating Christmas parties, awards ceremonies, bucks nights, weddings and various festivals. Antiques tourism has become another niche where rural and regional towns have made notable inroads (Michael 2002). Based around antiques stores, bric-a-brac and second-hand markets, charity shops and collectors fairs – all of which are more common in country towns than in big cities – antiques tourism is an altogether unobtrusive option, unlikely to create local opposition, and requiring little in the way of infrastructure investment.

Farm tourism has similarly changed. Whereas traditional farm-stay tourism attracts middle-class families from cities, keen to expose children to farm life and the origins of food (Kidd *et al.* 2004), increasingly popular are schemes attracting younger backpackers and students who might otherwise steer clear of rural places. Over 2,400 Australian farms now take WWOOFers – Willing Workers on Organic Farms – who volunteer for 4 to 6 hours per day in exchange for meals and accommodation (hence the appeal to 'Get out of the Tourist scene and taste our culture first hand' (http://www.wwoof.com. au/). Rural tourism encompasses active participation, and blurs the boundaries between 'work' and 'leisure'.

Sporting competitions and activities can be lucrative. As Australians become more and more interested in cycling (after vicariously consuming the idyllic rural landscapes of the Tour de France), an increasingly sophisticated national circuit of road races has grown – most conducted in rural regions. Cycle tourism has emerged as a focus of regional tourism policy development and formed the basis of a specialist touring niche (Lamont 2009). There are now Tours of Toowoomba (QLD), Gippsland, Great South Coast, King Valley and the Murray

River (Victoria), and Mersey Valley (Tasmania); while longer-established races such as the Grafton to Inverell Cycle Classic, first run in 1961, have gained greater publicity and become more formal elements of regional tourism promotion. In rural Victoria, as in New Zealand and the United Kingdom, disused railway lines have been converted to cycle trails, connecting rural places on flat paths that were once rail tracks, promising 'the perfect pedal to produce adventure' (http://www.murraytomountains.com.au/the-rail-trail/). No less than seven specialist cycle tour operators operate on the Murray to Mountains trail between Wangaratta and Bright. It also helps that cycle tourists tend to be professional, well-educated and highly-paid.

A different kind of mobile touring demographic are 'grey nomads', people 'aged over 50 years, who adopt an extended period of travel independently within their own country. They travel by caravan, motor-home, campervan or converted bus for at least three months, but often up to several years, moving around Australia' (Onyx and Leonard 2005: 61). Unlike many other tourist segments, the grey nomad scene continues to expand: in 2012 grey nomad tourism had grown 90 percent in a decade (compare with Figure 10.1 for overall tourism numbers), and was estimated to have generated over 10,000 jobs, 90 percent of them outside capital cities (*The Grey Nomad Times* 2012). Over 400,000 registered RVs (recreational vehicles) exist in Australia, and in Queensland alone over 80,000 travel annually, spending on average $700 per week (Mackander 2012). Grey nomads have generated significant discussion in regional development circles because they tend to travel routes avoided by other tourists, especially inland; stay in small and remote places; spend longer on the road; and although they do visit high profile tourist attractions are also happy to visit towns with more prosaic offerings, sampling 'everyday life' in the bush, from local markets to fishing and bird watching (Prideaux and McClymont 2006).

Some places have been more effective than others in tapping into this burgeoning market. Exactly which towns benefit most is influenced by a combination of beneficial location on trans-continental highways, underlying tourism attractiveness, openness of the town to older visitors, and provision of available facilities. Grey nomads follow a well-trodden route (word of mouth is very strong) and look for towns with affordable rest stop parks. Unlike American snowbirds, many Australian grey nomads detest permanent caravan parks, and are 'motivated by a different set of intentions... vehemently opposed to staying in an organised resort of any sort' (Onyx and Leonard 2005:61). Traditional caravan parks have become increasingly expensive in recent years, with facilities geared more towards young families with children (such as waterslides and playgrounds). Instead, grey nomads prefer cheaper open-space areas with minimal fuss (so long as there is access to sewage dump facilities), and often seek out remote camping grounds in national parks rather than urban locations. The Mary River in the Kimberley region and Gregory Downs in northwest Queensland are two such popular examples (Onyx and Leonard 2005).

Online forums and dedicated magazines and publications, including those of the influential Campervan and Motorhome Club of Australia, rate towns and campsites, according to their facilities for grey nomads (see for example, http://thegreynomads.activeboard.com/forum.spark?forumID=51933). Some regions, such as Queensland's Sunshine Coast, suffer from reputations as too expensive and unwelcoming of grey nomads, favouring a younger tourism market instead. Goondiwindi in southern Queensland is ideally located on a major east coast route, and is a popular fishing spot, and accordingly sees considerable through-traffic of grey nomads. But the expense of its local caravan park, and the difficulty of finding parking for large caravans close to supermarkets in the CBD have led grey nomads to stay there for one night only, or to by-pass altogether, meaning significant missed revenues for the town (Prideaux and McClymont 2006, Bentley 2012). Transitioning from a traditional agricultural base towards tourism thus requires not just attractions, accommodation and suitably skilled local people, but an increasingly specialised approach, with targeted marketing and integrated planning.

Festival Spaces and Places

Nowhere have the imperatives to specialise and diversify been more apparent than in the staging of festivals and events. Large scale gatherings are nothing new in rural Australia – agricultural shows, rodeos and travelling boxing tents have been mainstays for over a century (Hicks 2001, Darian-Smith 2011). From the 1980s onwards a multitude of more contemporary and specialist festivals had begun to appear in small towns, many for no special reason other than for entertainment and because local authorities wanted to enhance local cultural life. Others started because of desires to promote particular local industries, because of the efforts of enthusiastic individuals with a distinctive pastime or hobby, or because local creative artists, especially musicians, sought outlets for exhibition and performance where few previously existed. Many festivals in small towns happened just once, or lasted for a few years depending on the enthusiasm of the local organising committee or local government-employed festival organiser. Others survived, gained reputations and developed their own heritages and traditions, rising from small beginnings to national and even international prominence. The more well-known include the Tamworth Country Music Festival (and competitors in Gympie and Mildura), Deniliquin's Ute Muster (the world's largest annual gathering of pick-up/ 4WD vehicles), Maleny/Woodford Folk Festival, Byron Bay's East Coast Blues and Roots Festival, the Parkes Elvis Festival and Goulburn's Blues Festival.

Festivals are now both numerous, and kaleidoscopically diverse, throughout non-metropolitan Australia (Table 10.2). Gibson *et al.* (2010) tracked the extent and impact of festivals across non-metropolitan Tasmania, Victoria and New South Wales – identifying over 2800 festivals in a calendar year outside the

Table 10.2   Numbers of Festivals, by Type, Tasmania, Victoria and New South Wales, 2007

Type of festival	TAS	VIC	NSW	TOTAL**	% of total
Sport	86	485	488	1059	36.5
Community	45	216	175	436	15.0
Agriculture	19	146	215	380	13.1
Music	13	116	159	288	9.9
Arts	12	73	82	167	5.8
Other*	7	87	71	165	5.7
Food	10	53	67	130	4.5
Wine	7	49	32	88	3.0
Gardening	20	43	14	77	2.7
Culture	2	21	11	34	1.2
Environment	1	8	12	21	0.7
Heritage/historic	4	8	7	19	0.7
Children/Youth	0	10	5	15	0.5
Christmas/New Year	0	10	2	12	0.4
Total	**226**	**1325**	**1340**	**2891**	**100**

*   The other category includes small numbers of the following festival types: Lifestyle, Outdoor, Science, Religious, Seniors, Innovation, Education, Animals and Pets, Beer, Cars, Collectables, Craft, Air Shows, Dance, Theatre, Gay and Lesbian, Indigenous, and New Age.

** The total for this table is slightly more than the total number of festivals in the database, due to counting of some festivals in more than one category. This occurred when separating categories proved impossible (for example, for 'food and wine festivals')

Source: ARC Festivals database, 2007

capital cities of those states. Festivals generated an estimated 176,000 part-time and full-time paid jobs (whereas agriculture employed 136,000 people in the same non-metropolitan parts of those states). The most common were sporting, community, agricultural and music festivals – which together made up three-quarters of all festivals. Even though these festival types dominated, within these there was endless diversity; 'community' festivals covered everything from Grafton's historic Jacaranda Festival (named after the town's signature tree) to Kurrajong's Scarecrow Festival, Lithgow's Ironfest, Nimbin's Mardi Grass (a marijuana pro-legalisation festival), Ettalong's Psychic Festival, Tumut's Festival of the Falling Leaf, Myrtleford's curiously amalgamated Tobacco, Hops and Timber Festival, and Benalla's equally bizarre Wheelie

Bin Latin American Festival. Similarly varied were sports festivals, covering everything from fishing to billy carts, pigeon-racing, hang gliding, dragon boat racing and camp drafting.

Locations with the most festivals tend to be large regional towns (Ballarat, Mildura, Wagga Wagga, Launceston), regions reliant upon tourism industries (Snowy Mountains, Coffs Harbour, Surf Coast) or coastal 'lifestyle' regions with mixes of tourism and retiree in-migration (Lake Macquarie, Bass Coast, Great Lakes) (Table 10.3). Hotspots for festivals are within 'day tripper' driving distance of Sydney and Melbourne (e.g. Blue Mountains). Nevertheless, as Table 10.3 shows, many of the places with the highest festival-per-population scores were in inland areas, not necessarily known for tourism, or proximate to capital cities. Several towns and villages in the Riverina region (surrounding the Murray River in NSW and Victoria) have become festival places, including Wakool, Narrandera, Holbrook and Jerilderie. Such patterns are a function of high levels of local entrepreneurialism and social capital, good available transportation, extent of orientation to the tourism industry, and regional 'contagion' effects (where neighbouring towns gain inspiration or steal ideas from each other).

The vast majority of festivals are run by non-profit organisations, usually tiny in size. Only 3 percent of the festivals surveyed were run by private sector/profit-seeking companies (Gibson *et al.* 2010). The usual aims of festivals are linked to the pastimes, passions or pursuits of individuals on organising committees, or to socially- or culturally-orientated ends such as building community, rather than direct income-generation. New livelihoods have emerged within regional market circuits, notably in and around Byron Bay (NSW) and Denmark (WA), where success is dependent on cashing in on the social and cultural capital brought by new migrants, local artists and 'alternative' subcultures (Curry *et al.* 2009). Such markets feature a mix of local organic produce, arts and crafts, bric-a-brac and vintage clothing, and provide incomes to local farmers and small-scale enterprises, and organisations such as schools and bushfire brigades. Meanwhile the broader festival logistics and stallholder circuits generate employment through companies hiring tents, portaloos and PA equipment, as well as the seemingly ubiquitous kebab, gozleme and chorizo sausage stallholders.

Although there are costs and risks, festivals are an under-acknowledged and yet potentially significant component of strategies to develop the post-productivist countryside. Festivals may be more or less lucrative in terms of total monetary gains; but cumulatively – from their sheer ubiquity and proliferation – they diversify local economies, insure against market fluctuations (because they are mostly non-profit and involve non-monetary transactions and resource-sharing) and improve local networks, connecting volunteers, diverse paid workers and local institutions. They also frequently advance laudable goals of inclusion, community and celebration.

Some small towns have been so successful that festivals have substantially contributed to economic and social development, and resulted in those places

Table 10.3 Festival Capitals: Top 20 LGAs by Number of Festivals, and Festivals per 10,000 Population, Tasmania, Victoria and New South Wales, 2007 (based on 2006 census data population counts)

Top 20 by number of festivals				Top 20 by festivals per 10,000 population			
LGA	Population	Festivals	Festivals per 10,000 population	LGA	Population	Festivals	Festivals per 10,000 population
Ballarat (Vic)	80045	73	9.12	Wakool (NSW)	4807	22	45.77
Snowy River (NSW)	18737	62	33.09	Towong (Vic)	5972	23	38.51
Alpine (Vic)	17806	60	33.70	Buloke (Vic)	6982	26	37.24
Greater Taree (NSW)	42943	55	12.81	Narrandera (NSW)	6485	22	33.92
Lake Macquarie (NSW)	177619	54	3.04	Tumbarumba (NSW)	3551	12	33.79
Greater Geelong (Vic)	184331	52	2.82	Alpine (Vic)	17806	60	33.70
Wollongong (NSW)	181612	50	2.75	Snowy River (NSW)	18737	62	33.09
Hobart (Tas)	47321	48	10.14	Barraba (NSW)	2138	7	32.74
Mildura (Vic)	48386	44	9.09	Queenscliff (Vic)	3078	10	32.49
Warrnambool (Vic)	28754	43	14.95	Bombala (NSW)	2469	8	32.40
Greater Bendigo (Vic)	86066	43	5.00	Holbrook (NSW)	2340	7	29.91
Delatite (Vic)	21833	42	19.24	King Island (Tas)	1688	5	29.62
Shoalhaven (NSW)	83546	40	4.79	Nundle (NSW)	1351	4	29.61
Wagga Wagga (NSW)	55058	36	6.54	Strathbogie (Vic)	9171	24	26.17
Surf Coast (Vic)	19628	36	18.34	Merriwa (NSW)	2337	6	25.67
Greater Shepparton (Vic)	55210	35	6.34	Pyrenees (Vic)	6360	16	25.16
Bass Coast (Vic)	24076	34	14.12	Tallaganda (NSW)	2637	6	22.75
Great Lakes (NSW)	31388	33	10.51	Jerilderie (NSW)	1786	4	22.40
East Gippsland (Vic)	38028	33	8.68	Coolah (NSW)	3682	8	21.73
Mount Alexander (Vic)	16174	31	19.17	Ararat (Vic)	11102	24	21.62

Source: ARC Festivals Project database, 2007

becoming better known because of their festivals. Meredith (Victoria) was once a sleepy highway town of 1,100 people until three friends decided to throw a gig for alternative bands on a parents' family farm, 12 kilometres away. In subsequent years the Meredith Music Festival, held every December, became renowned for alternative rock music, its quasi-anarchic rural atmosphere, laid-back security and non-commercial format. According to one founder, Greg Peele,

> When we started Meredith there was no other festival like it. We had no real model to work with. It fell into place. Each year something new was added and there was never any forward plan. We didn't have a three-year strategy. We just said "what did we do poorly last year?" and one of the key aspects was we wanted to get better before getting bigger. It wasn't about getting more people and it wasn't about making lots of money. It didn't make money (quoted in Gibson and Connell 2012: 21).

But it did bring vital incomes to local caterers, petrol stations and the general store, and firmly established Meredith as a festival destination. From obscurity Meredith became an iconic place for alternative music fans. Likewise, Port Fairy (Victoria), once a tiny fishing and farming village, and now dependent on second-home owners and tourists, has been boosted annually by its folk festival, one of the most famous in Australia. That success had enabled it to build a series of other festivals, which have included a Koroit Irish Festival, Rhapsody in June (mainly for local music), a Spring Music Festival and a Singing Up Country Festival (of Aboriginal music). In Parkes its Elvis Festival succeeded in generating significant economic benefits, fostering a sense of community and gaining nationwide notoriety – all by associating with a dead performer who had never visited Australia, let alone Parkes (Brennan-Horley *et al.* 2007).

Festivals invite diverse encounters between incomers and local residents, between travelling artists or producers and local leaders and businesses. These encounters may alter the cultural dynamic of places, demand responses from local leadership, challenge local traditions, and trigger positive and negative reactions on both sides. Festivals are also a sphere of performativity, where cultural identities are constructed for and by people and places, and enacted throughout the duration of an event. Festivals may challenge or conform with norms, trigger conflicts, play a part in shifting notions of place identity, and include and exclude certain members of communities. Hence Curryfest in Woolgoolga (NSW) reflected not just the presence of that town's growing Sikh community, but enabled expansion of regional food cultures, and more broadly promoted multiculturalism (Milner and Hughes 2012). Not to everyone's tastes, Scone (NSW) hosted Metalstock (featuring bands such as Anarchsphere and Rake Sodomy), and Daylesford's ChillOut brought thousands of gay and lesbian tourists – much to the chagrin of more homophobic elements of the surrounding rural community (Gorman-Murray *et al.* 2008). Some festivals have encouraged communities to celebrate rather than eschew difference. Bellingen's Global Carnival epitomised multiculturalism; and

Nundle transformed its annual Gold Rush heritage festival into an annual Chinese themed festival (Khoo and Noonan 2011). Notions of rurality are rapidly changing and becoming increasingly diversified.

Unsurprisingly, many rural festivals have become linked to gastronomy and viticulture, and to a wider elite countryside. Hence gourmet food, wine and music have combined to shape a more Tuscan countryside in the Hunter Valley, a new Mediterranean has been hinted at in the olive growing areas of northern Victoria, and the Barossa has frequently reinvented and accentuated its German heritage (Peace 2011). The Hunter Valley first transformed itself into 'Wine Country' in the early 1990s, a process which subsequently involved further transformation and diversification into a gourmet region. In 1997 the Wine Country tourism brochure announced the inaugural Wyndham Wine and Food Festival (replete with 'the music of wandering songsters and musicians'), arguing that 'The Hunter Valley, renowned for its world famous wines and award winning restaurants is fast becoming renowned for its hedonistic indulgence each October at Jazz in the Vines'. Opera in the Vineyards was similarly established, and Jazz by the River took its place in the Upper Hunter.

By 2006 cheeses and olives were increasingly important (a Gourmet Cheese Festival had arrived), farmers' and growers' markets thrived, new festivals were established, some that also linked into cuisine (a Thanksgiving Festival, A Taste of Tuscany) and others that offered diversity (a Film Festival, a Rodeo, Steamfest) and multiple music festivals, including a Buskers Challenge (and others where music merely played a part, such as the St Patrick's Weekend Celebrations, the Festa del Vino and the Lovedale Long Lunch – 'wine and dine your way around 7 participating wineries over the weekend enjoying a glass of wine, gourmet food, fabulous music and art'). In certain selected parts of rural and regional Australia, local councils, vineries and restaurants were steadily transforming the countryside, in a degree of fortuitous collusion with magazines such as *Australian Gourmet Traveller* and *Country Style*, to fabricate a heritage that conveyed regional distinctiveness, a legacy of small farms, wineries and farmers markets, and an artisanal tradition that was to be celebrated in festivals.

Wineries in particular have become festival destinations, so much so that many now depend on concerts for their survival, and have even ripped out vines to create more space for audiences. At Bimbadgen Estate in the Hunter Valley:

> At the moment in Australia there's an oversupply of wines. So we pulled out vines for a good commercial reason: to increase the size of the amphitheatre. We can always buy grapes down the road somewhere (quoted in Gibson and Connell 2012: 7)

Bimbadgen Estate now has an amphitheatre capable of accommodating 8,000 people, and claims to 'epitomise Bimbadgen's philosophy – that it's not just all about the wine, but the experiences you have whilst enjoying it'. Probably the first concert in a winery was in 1985 when Leeuwin Estate (WA) hosted the

London Philharmonic Orchestra. In the 1990s, as wine and music increasingly came together, wineries took on broader roles that developed into wine and cheese tasting and then meals, accommodation, and subsequently concerts. A glut of wine on the Australian market for more than a decade has meant low prices and necessary diversification, hence the promotion of concerts and festivals. They have assumed and attracted a metropolitan population willing to travel some distance for a unique (and expensive) performance. Concerts therefore used the ambience of wine-producing regions such as the Hunter Valley, McLaren Vale and Margaret River, although some wineries developed particular genres, usually jazz and opera, or staged the 'return' of 'heritage' stars, such as Leonard Cohen and Elton John.

Wine concerts, although emblematic of post-productivism, are in some ways an exception. Most rural festivals remain small, community-orientated affairs where profits are low, but volunteerism and local multipliers are high. In exceptional cases, as at Deniliquin, festivals provide communities with a viable economic alternative to agriculture, and beer and rum are vastly more popular than wine. In many more less exceptional circumstances, festivals provide rural communities with some tourism revenue, but more importantly coping mechanisms at times of drought and economic hardship – catalysing community in the name of fun.

Rural Creativity

A final element of the post-productivist countryside is creativity. In the past decade much has been written about the geography of creative industries such as film, music, design and fashion. Creativity is said to be *the* salient feature of contemporary post-industrial capitalism, fuelling innovation and investment and therefore responsible for economic fortunes, as well as being a somewhat intangible quality in places ('the buzz' in cultural milieu) said to generate lifestyle-led in-migration (Florida 2002). Accordingly, municipal and state authorities in Australia, as elsewhere, have rushed to develop strategies aimed at branding places as creative, enhancing creative industry growth and appropriating the arts as regional development strategy. In rural Australia the creativity agenda has been appropriated to varying degrees, and with varying degrees of success.

Broken Hill, in far west NSW, has a history dominated by mining, but has since become one of the more prominent promoters of creative industries. It was where the corporate giant BHP (Broken Hill Proprietary) was incorporated; its silver, lead and zinc mine still towers over the town. Following decline in mining employment from the 1940s (a result of increasing capital intensity and mechanisation), the town's population contracted and exposure to the vagaries of international commodity prices became more pronounced. Diversification included an emerging tourism market in the 1970s and 1980s and increasing service industry employment (especially in health and community services, as it has become a 'retirement town'). Growing tourists created a market for local arts and crafts, and over time Broken Hill saw the evolution of a visual arts scene (Andersen 2010). Central were a group of artists known as the 'Brushmen of the Bush' – Pro Hart, Jack Absalom, Eric Minchin, John Pickup and Hugh Schulz – who from the 1970s painted 'outback art' full of arid landscapes, surreal figures and vivid skies. Since then a stable arts industry has evolved, linked in part to the passing tourist trade, but also with notably consistent demand for 'family presents, interior decoration, gifts from work colleagues for someone leaving town' (Andersen 2010: 80). It is 'a confident, growing sector of older artists who earn more income and have more time to work on creative practices than their metropolitan counterparts' (Andersen 2010: 74). Nevertheless 'factionalism, long-standing divisions, and professional rivalries hinder cooperative effort' (Andersen 2010: 79).

Beyond vernacular creative industries, far inland regions have played a prominent role as scenic locations for the metropolitan film industry. Broken Hill and nearby Silverton (population 50) provide ideal desert landscapes, reliably dry weather and suitably rustic colonial era buildings. More than 60 feature films and 200 commercials have been filmed in the Broken Hill region since the 1960s, including *Priscilla: Queen of the Desert* and *Mad Max 2* (Andersen 2010). Red Dog was set in Western Australia's Pilbara region; Baz Luhrmann's epic *Australia* was filmed in Kununurra; *Rabbit Proof Fence* was shot in various parts of outback South and Western Australia; *The Proposition* in Winton; and *Bran Nue Dae* in Broome (the home town of its musical authors, the Pigram Brothers). Nevertheless, some places are just too remote and expensive. Prohibitive costs meant that some of the filming of *Red Dog* took place instead only 40 kilometres outside Adelaide, at a place called Middle Beach, a location where other feature films seeking an 'outback' locale, including *Australian Rules* and *Wolf Creek*, were also shot.

Remoteness and isolation remain perennial challenges (Luckman 2012). In Darwin, challenges of remoteness included immediate issues such as the costs associated with flying in performers and artists for festivals and exhibitions, and more complex issues such as the difficulties of accessing key industry gatekeepers (Gibson *et al.* 2012). Musicians growing up and honing skills in rural and regional contexts know that career development usually requires relocation to larger centres, if not overseas, to music industry hubs. In Broken Hill remoteness meant 'limited types of creative making; wariness of newcomers and new ideas; the loss of young people; limited access to business expertise, production services and training; lack of cultural stimulation; and high transport costs' (Andersen 2010: 84). On the flip-side, isolation brought with it 'local landscape and culture; a friendly community; lower-than-city costs for accommodation and studio space; freedom from city-based art 'fads', stress and busyness; and a 'quality of life', time and a clear view' (Andersen 2010: 84).

New capacities in information technologies provide some measure of overcoming the difficulties of distance. Indeed, for many rural workers, teleworking is increasingly their only viable option (Simpson *et al.* 2003). For indigenous hip-hop musicians in the Torres Strait new telecommunications and recording technologies have helped counteract difficulties of being remote from key creative centres (Warren and Evitt 2010). Music is recorded using free software, often on communally provided computer equipment, and tracks are uploaded onto MySpace and You Tube for distribution. And yet distance, in combination with cultural norms about what constitutes 'authentic' indigenous creative expression, continues to circumscribe paid performance opportunities. Indigenous hip-hop musicians rely on gigs at national indigenous festivals and at events celebrating indigenous culture – but fail to secure slots on 'mainstream' festival line-ups in cities.

More generally, use of information technologies to overcome problems of distance is uneven, shaped by availability and speed of connection, and the multiple social and economic uses to which such technologies are put. Geography still matters, deeply circumscribing digital practices (Gibson *et al.* 2012). Even where internet connections are possible they can be too slow to make transactions and social uses meaningful; or, in the case of remote Aboriginal communities, such functions as online banking and dealing with government agencies online are exclusively in English, and are overly complicated, so their potential has not been maximised (Thomas and Rennie 2012). Similar problems constrain the development of telehealth in remote areas, and so make rural residence more difficult (Wakerman *et al.* 2008). Rural people 'make do' with what mobile and internet technologies are available, but much more could be done to enable fuller participation.

There are other missed opportunities too. Although Tenterfield has attempted to make something of it being the birthplace of singer Peter Allen, Wangaratta has done little with the fact that it is was where Nick Cave grew up. At Goulburn's Big Merino its souvenir shop stocks an impressive range of knitted goods, but in early 2013 the only available Australian-made knitted products were socks (from Tasmania). Everything else was made in New Zealand, surprising since knitting was once a ubiquitous type of rural creativity, the bastion of many a country fair or school fete. In Ravensthorpe (Western Australia) one basic problem was that there were not even postcards of the town available for sale locally, spurring local community groups and artists to make their own (Mayes 2010) – creativity by necessity.

Finally, rural areas provide opportunities for articulations of diverse and challenging creativity. Indigenous rural and remote creativity is more than mere means to economic development (though it does bring some measure of that); but rather an avenue to cultural maintenance, empowerment and expression, including articulation of vocal criticism and protest against the impacts of colonialism (Gibson and Connell 2004). Arts programs are often first and foremost a means to social participation (Anwar-McHenry 2011). Woolgoola Curryfest provides opportunities for community creativity and social cohesion between Sikhs and others; transcending the overly white, urban vision often promoted by creative city boosters such as Richard Florida. In Bega meanwhile – a staid town otherwise famous for its traditional dairy industry, notably cheese – the Spiral Gallery was established in the 1980s as a space for women to express their artistic identities.

In time, it surpassed the comparatively stuffy regional art gallery as a site of 'edgy' artworks, but moreover enabled women to explore their personal identities outside traditional gender roles within which they otherwise felt circumscribed (Waitt and Gibson 2013). For some women this low-key, low-cost piece of cultural infrastructure even became a kind of 'life-support system', providing a sympathetic social network for those escaping drugs, marriage disintegration and domestic abuse.

In that same town, the perennial struggle to keep a commercial music scene going gave rise instead to a vibrant community music scene, focused on choirs. Choirs were non-profit and closely tied to local high schools and music teachers; they generated virtually no commercial benefit, but enabled widespread creative involvement from all ages, backgrounds and both genders, virtually for free, and in the process overcame isolation, and provided an instant emotional support mechanism (Gordon 2012). Some important forms of creativity remain 'hidden' in small, marginal and remote places without obvious hubs or creative districts.

Conclusion

Agriculture still matters – it makes up about 6 percent of employment (Table 10.1) – and remains a significant export sector. Moreover farming provides powerful place identities and associations for rural Australian regions – an important heritage that is unlikely to be jettisoned without dissent. Indeed to some extent it has diversified to meet the needs of markets and tourists. However, when 94 percent of the workforce are employed in other industries, the future requires diverse and heterogeneous approaches. Services, tourism and creative industries will be part of that mix.

There are nevertheless problems in overstating the simplicity or likely significance of the post-productivist countryside. Marketing places is complex, can often fail, and runs the risk of upsetting local people who hold different views of how to represent their community. Rural tourism too has its limits. Farmstay tourism works best when farm owners have time for visitors, involve them in daily activities and entertain visitors' children – but this places new demands on already busy farming people, some of whom may not automatically shine to visitors, or who may tire of intrusions. Not all farmers are open to their daily physical work transmuting into 'a managed and professional service encounter' (Kidd *et al.* 2004:61). If promoting the post-productivist countryside is the goal, then part of that needs to be finding ways to put together enterprises that suit the aspirations and capacities of local people.

In many ways the prospects for tourism commodification remain limited, and competition is intense and ever-evolving. The more places launch wine and food fairs, jazz and blues festivals or promote farmstay tourism, the less novel they appear. All over Australia agritourism has been discussed as a potential farm diversification activity, but for a range of reasons that include remoteness and lack of capital, it has not been overwhelmingly successful, and in many parts remains unviable (Fisher 2006). The constraints of distance are still palpable, and although digital, mobile and wireless technologies have enabled new kinds of business and creative practices, there remain digital divides along geographic and demographic lines. In western NSW the $1 farmhouse rental scheme did gain good media coverage, but could not address basic problems of isolation, lack of services, and poor employment prospects in tiny, remote places.

Amenity has emerged as a leitmotif for post-productivism and a driver of population flows. Even if lifestyle, culture and rurality are intangible, they are ingredients in the regional development equation to be taken seriously. Yet, 'in the high amenity communities in which ... migrants are making their new homes, local demographic, socio-economic and land use structures are undergoing dramatic change, but not always along easily predicted lines' (Argent *et al.* 2010: 23). Local residents may resent any kind of rapid change. In a few places tourism and creative industries have been so successful that the emphasis is on stemming the tide of incomers to protect local tranquillity and quality of life, as in Noosa (Queensland), Byron Bay (NSW) and Bridgetown (WA) (Tonts and Greive 2002). Some places might have all the 'right' ingredients – location, natural attractions, a vibrant small-business sector, a willingness to promote tourism, good grass-roots creative scenes – and yet lack key local leadership, the 'glue' needed to hold the post-productivist economy together. Albury-Wodonga, on the NSW-Victorian border, has been cited as one such example lacking this critical 'soft' infrastructure (Jackson and Murphy 2006). Other places may simply become too popular and through an 'unfettered pattern of development' (Tonts and Greive 2002: 58) destroy the very things – scenic rural landscapes, quietude, gentility, friendliness – that attracted people in the first place.

Even gradual change eventually brings with it new challenges. Flower shows, once a stronghold of country women's associations and a link to local cultivation, faded in places like Gympie because 'an ageing population, an inability to attract new members due to the increasing number of women in the workforce and the changing aspirations of women themselves, as they sought to move away from their traditional, gender-prescribed roles into more active positions in society' (Edwards 2011: 99). Agricultural shows have slowly contracted and amalgamated, and eventually had to confront issues of ongoing viability in regions with declining farming workforces. Some embraced the 'modern' world of rides and showbags, morphing into entertainment events, while others such as at Walcha (NSW) went 'back to their roots' and tightly controlled their offerings to present an 'authentic' community event. Either way they eventually confronted the need to update and focus on a strategy.

For rural places seeking to turn around decline or diversify into post-productivist activities, perhaps the hardest task of all is to create from scratch the emotional attachments, the sense of belonging to a rural place that sustains populations, and attracts outsiders repeatedly. It is possible, as in its small way the very successful home-hosting scheme at the Parkes Elvis Revival

Festival demonstrates (Li and Connell 2012). But such efforts take patience, generosity, appropriate structures (with a degree of hands-off support from local government) and ultimately good leadership (Onyx and Leonard 2010). Exactly how 'post' the post-productivist countryside in Australia becomes is impossible to gauge. What is certain is that leisure, tourism and creative industries will be a part of it, and a part that will generate a distinctly different, evolving rural and regional Australia.

References

Andersen, L., 2010, 'Magic Light, Silver City: The Business of Culture in Broken Hill', *Australian Geographer*, 41: 71-85.

Anwar-McHenry, J., 2011, 'Rural Empowerment Through the Arts: The Role of the Arts in Civic and Social Participation in the Mid West Region of Western Australia', *Journal of Rural Studies*, 27: 71-85.

Argent, N., Tonts, M., Jones, R. and Holmes, J., 2010, 'Amenity-led Migration in Rural Australia: A New Driver of Local Demographic and Environmental Change?', in G.W. Luck, R. Black and D. Race (eds), *Demographic Change in Australia's Rural Landscapes*, Dordrecht: Springer, 23-44.

Arthur, C., 2011, 'Giant Ant to Put Outback Qld "on the map"', *ABC News*, 30 March, http://www.abc.net.au/news/stories/2011/03/30/3177378.htm?site=northwest.

Bentley, B., 2012, 'Wake up "Gundy" – You're Missing Out', *Goondiwindi Argus*, 5 December, http://www.goondiwindiargus.com.au/story/1166263/wake-up-gundy-youre-missing-out/.

Botterill, L., 2006, 'Soap Operas, Cenotaphs and Sacred Cows: Countrymindedness and Rural Policy Debate', *Public Policy*, 1: 23-36.

Brennan-Horley, C., Connell, J. and Gibson, C., 2007, 'The Parkes Elvis Revival Festival: Economic Development and Contested Place Identities in Rural Australia', *Geographical Research*, 45: 71-84.

Clark, D., 2004, *Big Things. Australia's Amazing Roadside Attractions*, Melbourne: Penguin.

Connell, J. and Rugendyke, B., 2010, 'Creating an Authentic Tourist Site? The Australian Standing Stones, Glen Innes', *Australian Geographer*, 41: 87-100.

Connell, J. and McManus, P., 2011, *Rural Revival: Place Marketing, Tree Change and Regional Migration in Australia*, Farnham: Ashgate Publishing.

Costello, L., 2007, 'Going Bush: The Implications of Urban-rural Migration', *Geographical Research*, 45: 85-94.

Curry, G., Fold, N., Jones, R. and Selwood, J., 2009, 'Cashing In on Resources, Social and Cultural Capital: The Role of Local Markets in the Great Southern District of Western Australia', *Prairie Perspectives*, 11: 173-194.

Curtis, R., 2010, 'Australia's Capital of Jazz? The (Re)creation of Place, Music and Community at the Wangaratta Jazz Festival', *Australian Geographer*, 41: 101-116.

Darian-Smith, K., 2011, 'Histories of Agricultural Shows and Rural Festivals in Australia', in C. Gibson and J. Connell (eds), *Festival Places: Revitalising Rural Australia*, Bristol: Channel View, 25-43.

Destination NSW, 2012, *Regional NSW Time Series Statistics* – at June 2012, http://archive.tourism.nsw.gov.au/Regional_Tourism_Statistics_p625.aspx.

Edwards, R., 2011, 'Arranging Society with Flowers: The Rise and Fall of Flower Shows in Gympie, 1880-2004', *Journal of Australian Studies*, 35: 99-112.

Fisher, D., 2006, 'The Potential for Rural Heritage Tourism in the Clarence Valley of Northern New South Wales', *Australian Geographer*, 37: 411-424.

Florida, R., 2002, *The Rise of the Creative Class*, New York: Basic Books.

Frost, W., 2004, 'Heritage Tourism on Australia's Asian Shore: A Case Study of Pearl Luggers, Broome', *Asia Pacific Journal of Tourism Research*, 9: 281-291.

Gibson, C., 2013, 'Welcome to Bogan-ville: Reframing Class and Place Through Humour', *Journal of Australian Studies*, 37: 62-75.

Gibson, C. and Connell, J., 2004, 'Cultural Industry Production in Remote Places: Indigenous Popular Music in Australia', in D. Power and A. Scott (eds), *The Cultural Industries and The Production of Culture*, London and New York: Routledge, 243-58.

Gibson, C. and Connell, J., 2012, *Music Festivals and Regional Development in Australia*, Farnham: Ashgate Publishing.

Gibson, C., Luckman, S. and Brennan-Horley, C., 2012, '(Putting) Mobile Technologies in Their Place: A Geographical Perspective', in R. Wilken and G. Goggin (eds), *Mobile Technologies and Place*, New York: Routledge, 123-139.

Gibson, C., Waitt, G., Walmsley, J. and Connell, J., 2010, 'Cultural Festivals and Economic Development in Regional Australia', *Journal of Planning Education and Research*, 29: 280-293.

Gordon, A., 2012, *Community Music, Place and Belonging in the Bega Valley, NSW, Australia*, MSc Thesis, University of Wollongong.

Gorman-Murray, A., Waitt, G. and Gibson, C., 2008, 'A Queer Country? A Case Study of the Politics of Gay/Lesbian Belonging in an Australian Country Town', *Australian Geographer*, 39: 171-191.

Gorman-Murray, A., Waitt, G. and Gibson, C., 2012, 'Chilling Out in "Cosmopolitan Country"? Urban/Rural Hybridity and the Construction of Daylesford as a "Lesbian and Gay Rural Idyll"', *Journal of Rural Studies*, 28: 69-79.

Hess, K. and Waller, L., 2012, 'The Snowtown We Know and Love: Small Newspapers and Heinous Crimes', *Rural Society*, 21: 116-125.

Hicks, J., 2001, *Australian Cowboys, Roughriders and Rodeos*, Rockhampton: Central Queensland University Press.

Higgs, P., Cunningham, S. and Pagan, J., 2007, 'Australia's Creative Economy: Basic Evidence on Size, Growth, Income and Employment', Brisbane: QUT, Technical Report, Centre for Creative Industries. http://eprints.qut.edu.au/8241/1/8241.pdf.

Hooper, K. and van Zyl, M., 2011, 'Australia's Tourism Industry', *Reserve Bank of Australia Bulletin*, December Quarter, 23-32.

Hughes, M. and Macbeth, J., 2005, 'Can a Niche-Market Captive-Wildlife Facility Place a Low-Profile Region on the Tourist Map? An Example from Western Australia', *Tourism Geographies*, 7: 424-443.

Jackson, J. and Murphy, P., 2006 'Clusters in Regional Tourism: An Australian Case', *Annals of Tourism Research*, 33: 1018-1035.

Kennedy, M., 2011, 'Binding a Sustainable Future: Book Towns, Themed Place-branding and Rural Renewal – A Case Study of Clunes "Back to Booktown"', in J. Martin and T. Budge (eds), *The Sustainability of Australia's Country Towns: Renewal, Renaissance and Resilience*, Ballarat: Victorian University Regional Research Network Press, 207-226.

Kennett, H., 2012, 'Tourists Snap Up Grisly Snowtown Souvenirs', *Herald Sun*, 15 July, http://www.heraldsun.com.au/ipad/tourists-snap-up-souvenirs-of-snowtowns-past/story-fnbzs1v0-1226426458829.

Khamis, S., 2007, 'Gourmet and Green. The Branding of King Island', *Shima*, 1: 14-29.

Khoo, T. and Noonan, R., 2011, 'Going for Gold: Creating a Chinese Heritage Festival in Nundle', New South Wales, *Continuum: Journal of Media & Cultural Studies*, 25: 491-502.

Kidd, J., King, B. and Whitelaw, P., 2004, 'A Profile of Farmstay Visitors in Victoria, Australia and Preliminary Activity-based Segmentation', *Journal of Hospitality & Leisure Marketing*, 11, 45-64.

Lamont, M., 2009, *Independent Bicycle Tourism in Australia: A Whole Tourism Systems Analysis*, PhD thesis, Lismore: Southern Cross University.

Li, J. and Connell, J., 2012, 'At Home with Elvis: Home Hosting at the Parkes Elvis Festival', *Hospitality & Society*, 1: 189-201.

Luckman, S., 2012, *Locating Cultural Work: The Politics and Poetics of Rural, Regional and Remote Creativity*, Basingstoke: Palgrave Macmillan.

Mackander, M., 2012, 'Millions go begging', *Sunshine Coast Daily*, 4 November, http://sunshinecoastdaily.com.au/news/nomads-caravan-just-moves-on/1608332/.

Mayes, R., 2010, 'Postcards from Somewhere: "Marginal" Cultural Production, Creativity and Community', *Australian Geographer*, 41: 11-24.

Michael, E., 2002, 'Antiques and Tourism in Australia', *Tourism Management*, 23: 117-125.

Milner, L. and Hughes, M., 2012, 'From Bananas to Biryani: The Creation of Woolgoolga Curryfest as an Expression of Community', *Locale: The Australasian-Pacific Journal of Regional Food Studies*, 2: 119-139.

Onyx, J. and Leonard, R., 2005, 'Australian Grey Nomads and American Snowbirds: Similarities and Differences', *The Journal of Tourism Studies*, 16: 61-68.

Onyx, J. and Leonard, R., 2010, 'The Conversion of Social Capital into Community Development: An Intervention in Australia's Outback', *International Journal of Urban and Regional Research*, 34: 381-97.

Peace, A., 2011, 'Barossa Dreaming: Imagining Place and Constituting Cuisine in Contemporary Australia', *Anthropological Forum*, 21: 23-42.

Prideaux, B. and McClymont, H., 2006, 'The Changing Profile of Caravanners in Australia', *International Journal of Tourism Research*, 8: 45-58.

Rofe, M., 2013, 'Considering the Limits of Rural Place Making Opportunities: Rural Dystopias and Dark Tourism', *Landscape Research*, 38: 262-272.

Simpson, L., Daws, L., Pini, B. and Wood, L., 2003, 'Rural Telework: Case Studies from the Australian Outback', *New Technology, Work and Employment*, 18, 115-126.

The Grey Nomad Times, 2012, 'Grey Dollar Makes a Difference in the Bush', *The Grey Nomad Times*, 17 (30 March), 1.

Thomas, J. and Rennie, E., 2012, 'Nobody Uses the Internet Because the Government Says they Should', *Inside Story*, http://inside.org.au (accessed 18/10/12).

Tonts, M. and Greive, S., 2002, 'Commodification and Creative Destruction in the Australian Rural Landscape: The Case of Bridgetown, Western Australia', *Geographical Research*, 40: 58-70.

Waitt, G. and Gibson, C., 2013, 'The Spiral Gallery: Non-market Creativity and Belonging in an Australian Country Town', *Journal of Rural Studies*, 30: 75-85.

Wakerman, J., Humphrey, J., Wells, R., Kuipers, P., Entwistle, P. and Jones, J., 2008, 'Primary Health Care Delivery Models in Rural and Remote Australia', *BMC Health Services Research*, 8.

White, R., 2012, 'From the Majestic to the Mundane: Democracy, Sophistication and History Among the Mineral Spas of Australia', *Journal of Tourism History*, 4: 85-108.

Chapter 11

Soft Country? Rural and Regional Australia in *Country Style*

John Connell

> Each month *Country Style* presents inspirational people and interesting places
> from around Australia and samples all the good things our country has to offer
> (*Country Style*, June 2011, p. 14).

Many of the chapters in this book point to difficulties afflicting rural and regional
Australia – notably challenges to agricultural production, conflicts between
mining and agriculture, inadequate access to water and social services and
declining populations, marked by the out-migration of both the young and the
elderly. Rural and regional Australia can be a troubled place of social problems –
crime and isolation, of economic decline and natural hazard. Indeed various local,
state and private organisations have sought to develop strategies to retain rural
populations and encourage urban-rural migration. Such initiatives have included
regional growth centres, the efforts of Country Week (now the Foundation for
Regional Development) in NSW and Queensland, rentafarmhouse (which rents
unused farmhouses at peppercorn rents), Evocities and various others (see Chapter
5). Yet at the same time as a range of organisations have sought to discourage
outmigration, encourage return and attract new residents, sometimes seemingly
against the odds, a proliferation of lifestyle magazines have painted a very
different picture of verdant, 'cultured' middle-class rural landscapes with obvious
attractions.

In the past quarter of a century rural and regional Australia has been
reconfigured in a number of lifestyle magazines, including the glossy monthly
Country Style. It presents regional Australia as a place of nostalgia and return,
of heritage (homes, gardens, farmhouses and furniture), of virtue and morality,
devotedly restored, peaceful and close to nature, and of invisible modernity. It is
a post-productivist countryside of rolling green hills, not the 'boundless plains' of
sheep and wheat Australia, where the landscape and especially home gardens have
been gently tamed and nourished, and in turn nourish the residents. It is a place for
nurturing children, animals and plants. Production is of handicrafts and gourmet
food and wine, homemade and handmade, for farmers markets, boutique stores
and local festivals. Residents live in villages, cottages and farmhouses, rather
than towns, located on hillsides with views, often not far from the coast, near

Figure 11.1   *Country Style*

Source: *Country Style*, June 2011, reproduced with permission

certain idyllic 'hotspots'. Community and neighbourliness are legendary while cosmopolitanism exists only in food. *Country Style* embodies the performance of a new rurality. This country offers a sense of space and scale, a place for children and families and a degree of tranquillity: the antithesis of metropolitan life – the 'urban Babylon' of danger, pollution, noise, congestion, lack of community and so on – and one that is more urban myth and dreaming than the sometimes 'hard scrabble country' of alternative perceptions of rural Australia. While *Country Style* has no mission to encourage migration to rural Australia the superiority of rural life is absolutely implicit.

Country Style began life in 1988 as *Australian Country Style* until the 2000s when it became *Country Style*, though retaining its Australian focus. Through the years it has presented a determinedly and devotedly positive view of rural Australia, in a recurrent set of themes, through which it is possible to identify a contemporary countryside that embodies both notions of 'country' and 'style' and from which emerges a similar, but distinctly different, set of characteristics to those elaborated upon by Country Week (Connell and McManus 2011). These consistently emerge from the feature stories, the letter pages and editorials, and even from the adverts, embodied in the opening editorial excerpt. From a basic content analysis of *Country Style* from 2008 to 2012 nine broad but subtly distinct themes constantly recur: nature/environment; aesthetic agriculture; virtue/simplicity/morality; heritage; good food and wine; consumption; time; place/community/scale; and children and family. These are outlined below, as they are depicted in features and also letters.

Nature and Environment

Country Style Australia is without pollution of any kind, whether noise from traffic or aeroplanes, smells, or the visual pollution of advertisements and graffiti, but a place of peace and tranquillity, where the entirety of the landscape is to be appreciated and enjoyed. Implicitly, and explicitly (the notion of 'escape' recurs), the country is everything that the city is not. Indeed *Country Style* is something of a mantra against urban excess and distress; recurrent themes occur in the background of most stories:

> Earlier this year we moved from the concrete jungle of Sydney to a piece of paradise in north east Tasmania. Many months later I happily sit in my newly decorated parlour, gazing at the to-die-for view of green pastures and distant mountains, sipping tea and leafing through my *Country Style* magazines that I have collected from the very beginning. Ah yes … this is how life is meant to be! (CS August 2009)

> Being here is sanity…it's a lot easier to [work] here, amidst harmony, than dealing with traffic and peak hours (CS June 2011).

To say Melinda Trost's home is off the beaten track is no overstatement. Tucked in to the rainforest behind Byron Bay it is reached only after crossing causeways and gingerly edging down gravel roads deeply grooved by recent rains. Melinda and her partner live in a romantic cottage on the rambling hinterland property... The baby blue cottage is framed by a deep green tropical garden...Beyond the gate is an overgrown paddock that rolls down to a wide creek frontage – perfect for lazy summer swims or long winter walks. 'It looks like England doesn't it? We love that' says Melinda, looking down to the river and up to the rolling green hills opposite. 'Yet here we are part of a koala colony with echidnas and platypus' (CS June 2011).

Life at Tanah Merah displays many of the elements of that amiable rural idyll that poets and painters and writers have been feeding off for centuries. The approach to the homestead is through a ford over the creek and down a long dusty road straddled by paddocks full of sleek Hereford Cattle, and the wheat and oat crops that will soon make them even sleeker (CS June 2011).

The road traverses typical farming country, passing Lake Lyell and silent cattle ruminating on their good fortune to be in such picturesque paddocks (CS August 2011).

Overworked adjectives, as around Melinda Trost's property, exude pleasure. 'Green' hills are repeatedly 'rolling'. Aesthetic charms apply not only to the landscape but to houses and other buildings and even, in photographs, to decorative dogs, such as spaniels, rather than working dogs such as blue heelers. While landscapes may be 'dotted with lambs' they exclude snakes, flies and any noxious species.

Sometimes I stand and watch the sensational sunsets, while being deafened by birdsong, and I feel so very fortunate to have this life (CS August 2011).

Landscape and 'rolling hills' take on a particularly English quality. 'In the NSW Southern Highlands fine food complements views recalling rural England' (CS August 2008). Gardens especially have taken on Englishness in their organisation and seasonality. Kangaroos have seemingly been banished.

Aesthetic Agriculture

Rural landscapes are predominantly there to be gazed on, and to invoke desire. 'Driving through the open countryside around Rydal one day, an emerald paddock dotted with lambs stopped me in my tracks' (CS June 2011). Purchasing an 1860s cottage followed. Vistas are as likely to take in national parks as actively worked agricultural land. Nonetheless although agriculture is inescapable, only very exceptionally does it resemble large-scale wheat-sheep farming on featureless plains (with no elite or aesthetic status), but is almost always either small-scale, organic, slow food production, such as olives and grapes, oriented to farmers markets, embedded in concepts of paddock to plate, and with some 'heritage' value or targeting particular niche markets: a sort of post-productivist 'craft agriculture'.

Standard cows, sheep, pigs and grain hold no interest. Olives and avocadoes, and even coffee and quinoa, figure prominently. 'Working with chefs is a good way to get native ingredients like saltbush, sea celery and kunzea into the food vernacular' (CS September 2012). 'Most people focus on only one breed. I think there's only one other who has a bigger collection of rare [pig] breeds than me' (CS August 2011). 'The Dexters really are fascinating little cows ... the world's smallest breed. They all have different faces and personalities and they all have names' (CS June 2012). Animals too are adjectival: Belted Galloways and Droughtmasters, Isa Brown and Sussex chickens and every dog. The role of agriculture is that of creating a pleasant visual landscape, on limited acreage, rather than raising or resolving environmental issues. Rare excursions take in farms in the Upper Hunter – and wheat farms in Quirindi and Western Australia – so that standard agriculture has some place. 'He investigated drought-hardy sheep and settled on the Dorper, a South African breed that adapts well to harsh conditions' (CS June 2011). Similarly,

> After years of drought and now flooding, Australians are beginning to take heed of holistic farming ...Tarwyn Park is an oasis of regeneration in a region that appears in decline' (CS March 2011).

Nonetheless saffron growing in Tasmania, and similar featured crops, are at least as important. The actual work involved in agricultural production is rarely discussed. The reference to 'harsh conditions' is atypical, and drought is always in the past. Agricultural (and general) problems – such as floods and fires – are rarely evident and have been triumphed over.

> Not so long ago fire and floods ravaged this remote country property. They also brought love and marriage to its owners (CS March 2011).

Families exercise long-term stewardship over the land and manual labour, feeding farm animals, is more virtuous than mechanisation. In this sense *Country Style*'s agriculture is about improvement: careful use of 'slow water', nourishing the landscape rather than exploiting it and providing valuable experiences for children.

Country Style features a Farmers' Market of the Month, but that is usually as close as it comes to hinting at commercialism. At the markets produce is normally 'home made' or 'handmade' with the 'finest local' 'farm fresh' ingredients, sold in brown paper wrappings or even in wooden boxes. Markets also harbour rare foods such as 'tomatillos, purple peacock broccoli and purple dragon carrots' (CS September 2012). These are sold appropriately: 'The market has a policy of no plastic bags so be sure to bring your own shopping bags' (CS June 2011).

Farmers' markets moreover represent the regular celebration and performance of rurality, community and heritage skills, rather than an assembly of capitalist entrepreneurs. Markets are places of aesthetic pleasure for all the senses and the means of acquiring appropriate local food:

> At the Barossa Farmers' Market I love being able to talk to the people who have grown the produce I am buying, hearing their stories about the season or getting their serving suggestions. It can be the smallest of things, such as taking a moment to smell the liquorice scent of new season's fennel or watching the earth being shaken off baby carrots that really connects us with where our food comes from (CS June 2011).

The more virtuous the foods, the more laudable the consumers. Such obsessive excesses of foodism are neither unusual nor specific to *Country Style* (Poole 2012).

Local foods emphasise nature and seasonality: qualities missing from urban life. Fields merge with gardens.

> Beyond the mud-brick dwelling that faces north and overlooks soft hills and dense forest, there's a large vegetable garden, a chicken coop and the guest quarters that they call The Long Barn. Another barn houses Greg's studio, framing workshop and gallery. There's also an immaculate croquet lawn – and 52 sheep that call The Old Farm home (CS website).

Gardens, usually personal gardens, are a key part of the landscape and are both aesthetic and a source of some food. In *Country Style* they appear as productive as large scale agriculture. All houses naturally have gardens.

> Today it's packed with pretty perennials, many varieties of climbing and standard roses, rare bulbs, and deciduous trees, including silver birch, crab-apple and a spreading basket willow. 'Being out in the garden with the little wrens, finches and honey-eaters, watching the roses open – there's nothing more joyous' (CS June 2011).

> Rex and Dani's large garden could be called a pleasure ground, an 18th century term for an area filled with lawns, ornamental plantings and architecture. And the pleasures are many: a series of garden rooms and walks integrate native and exotic plantings, and there's a fine balance between the productive and the ornamental. 'The rhubarb is rampant and the garlic has crept in by stealth'. Most seasonal produce ends up in the kitchen. Dani runs a farmhouse style kitchen. Jar after jar of jams, chutneys and sauces line up with delicious home-baked cakes and scones.... One of the garden's most impressive areas is a water feature that uses a French Anduze pot as a centre piece. 'We were inspired to try this after we visited the Jardins des Marqueyssac in the Dordogne region of France' (CS June 2011).

Gardens are thus somewhat divorced from the Australian landscape in layout and content, where decorative and productive plants replicate more English gardens.

A somewhat gentrified elitist countryside, where barns have become residences and studios, with a minimalist, aesthetic – sometimes cute and un-intrusive – agriculture, where home gardens and their output are as important. Agriculture itself is also restorative and a 'heritage' activity, revitalising the land and regenerating particular exclusive animal species and unusual plants.

Virtue, Simplicity and Morality

Both in its aesthetic and moral qualities the countryside exudes simple virtues that are often associated with past times. A regular feature of *Country Style* is an article ('My Country Childhood') about a celebrity who now lives in the city but grew up in the countryside, experiences a personal sense of heritage, is nostalgic for its simpler pleasures and dreams of return:

> Even 18-year old Jessica Watson (of round the world by boat fame) was nostalgic
> about childhood – travelling through and camping in rural Queensland: 'as kids
> we were given a lot of freedom' (CS June 2011).

Christine Anu, Tim Flannery, Glenn McGrath, Thomas Keneally and Kerry O'Brien have all featured, invariably being nostalgic about the values that this engendered, the close relationships with grandparents and, occasionally, more distant ancestors, with community and with animals. Simply 'growing up on a farm gives you a real sense of independence and a good work ethic' (CS March 2012). A pastry cook who had returned from the city stated:

> Once I wanted to be big, but realised sometimes it's better to be a small fish. I
> don't want a big expensive life, I just want a life (CS June 2011).

More generally:

> It is hard to describe to people who have never experienced country life just
> how profound the peace, purity and organic nature of our lives can be. This
> is especially so if, like me, you don't have the eloquence of speech, or the
> appropriate adjectives to describe the 'food for the soul' that country life can
> provide (CS March 2011).

Country Style has established a competition for schools to prepare the Best Class or School Harvest table, to encourage children to plant and appreciate local food.

> Country Style, which has a strong interest in the 'paddock to plate' movement,
> is encouraging kids to step outside and explore the possibilities of the kitchen

garden. Stephanie Alexander [Kitchen Garden Foundation founder, cook and food writer] ... believes ... 'When children are encouraged to dig and plant, to water and pick, to chop and stir, to set tables with bunches of herb and flowers, to sit and pass platters to each other, they become more open to new flavours. It gives children enormous confidence as well, learning new skills, and provides an opportunity to learn patience and the pleasure of anticipation – versus instant gratification' (CS June 2011).

Implicitly this is impossible outside small rural communities.

Heritage

Heritage takes multiple forms but is centred on the built landscape and on the agricultural and related activities undertaken in the country. In *Country Style* people tend to live in farmhouses, or in grand or quaint houses and cottages, rather than in streets in towns. Occasionally residents occupy more interesting properties such as former barns, butter factories, jails and churches. *Country Style* has a regular wine writer, who doubles as producer of the Country Squire column, and 'lives in a former courthouse, police station and jail in the NSW Central West village of Cobbora near Dunedoo' (CS June 2011). Many such properties demanded and benefited from tasteful restoration, to create a rustic and historic feel. 'From the heritage-listed stables to the pictures on the walls, everything in this grand old grazing property comes with an intriguing history lesson' (CS July 2011). Restoration demands seeking out original timber floorboards, vintage windows and other heritage components.

Wendy Caird came home from France with a taste for small country towns... Before she purchased it the pretty weatherboard [in the Kangaroo Valley] had become unstable due to the busy burrowing of a wombat clan in residence beneath the floor. The house had been built in two stages. In 1880 it was a general store and then was occupied by the village tailor. [After restoration] what she has ended up with is a petit bijou ('little jewel') as the French would say (CS, April 2010).

The exterior is made almost entirely of recycled materials, including timber salvaged from wharves along the Murray River – "All kinds of ferries and paddleboats have probably been tethered to this wood over the years," Antony explains. The surfaces are roughly hewn and scarred, with remnant bolts trailing orange rust streaks down the wall. "There's something magical about the way ageing corrugated iron and timber reminds you of things you've seen in Australian landscapes your whole life, and that you associate with the bush in either the heat or the cold," Antony says. "At night in summer the house clicks and ticks as it cools down, and there's also the sound of the birds returning to the lake." (CS July 2011).

This is a home with heart, filled with treasures found, foraged, and made by hand. Incredible art line the walls, vintage rugs are strewn across the floors, and an old piano is played everyday. Nicola's niece once said that the house was like a giant cubby (CS April 2011).

Eight years ago my wife and I moved from the hustle of Brisbane to rural Queensland. Back then 'country style' meant an old farmhouse – broken down, mouldy and desperate for love, attention and a re-stumping. Today it means something completely different. With our own private parkland, panoramic views ... we have more time to reflect, read and write and have discovered that noticing beauty is as important as living within it (CS, July 2011).

Phrases and titles depict worthy and virtuous struggles. 'It was my aim to maintain the soul of the house' (CS August 2011), 'How a family saved an historic Mudgee farmhouse from Demolition' (CS May 2011) and texts emphasise the dignity of a very particular kind of labour.

"Every surface known to man had to be changed" Rosie says. Layers of carpet, wallpaper and "ghastly" linoleum were ripped out. Pressing on, the house was reclad and refurbished. New windows, French doors and skylights transformed the dark interior where original lining boards ...needed extensive renovation. The slab-laid stringybark floor was so warped that sanding was not an option...Wrap around verandahs and a new roof completed the project (CS June 2011).

"Its faded opulence" says Melinda, gesturing at the furniture and found objects that transform the cottage's simple white walls and floors. Doorways and windows have been replaced with ornate shutters found in northern India. Frames exhibit mirror samples with marble and gilded lace finishes. Bookshelves groan with tomes...Melinda's home offers a small oasis of opulence (CS June 2011).

It was 1997 when Elizabeth and Jim became the proud owners of the All Saints Anglican Church in the [Dandenong] mountain village of Ferny Creek. The quaint timber structure had been built in 1903 and then relocated several times round the Dandenong Ranges ... Creating a dream in such an undomestic space may seem daunting but for Elizabeth and Jim it was simply a dream come true... "I've travelled around the world, but I still think this is the most beautiful place" (CS June 2011).

Gardens too have been revitalised and restored. The most traditional of colonial landscapes becomes paradise regained, so creating a new sense of belonging:

> The garden renovation on this historic pastoral property would be instantly
> recognisable by the 19th Century pioneers ... By blending hard stonework with
> softer flowers and foliage, the Taylors have created a garden oasis around their
> homestead where past meets present in true harmony (CS July 2011).

Houses, once restored, have been properly (re)furnished, invariably with
collectables and antiques, to create the right ambience. *Country Style* has a regular
collectables writer, who also evaluates readers' 'precious objects'. Photographs
show and advertisements depict appropriate furniture such as 'luxury linen' (CS
June 2011) and the 'Cotswold furniture collection' (CS June 2011), that also
includes garden heritage:

> This Fermob '1900' side chair from Cotswold Furniture Collection features
> hand-forged scrollwork and is a pretty addition to any garden (CS June 2011).

and a host of other possibilities including 'made to order tables', 'heritage style
country gates', 'new and recycled furniture and homewares', backyard cabins',
'timeless design in pure natural fabrics', 'luxurious merino clothing', 'the Uralla
wool room', 'classic Adirondack chairs' and 'willie wildlife sculptures – the
world's finest birdbaths'. However real heritage items are infinitely superior:

> Inside, cherished pieces collected over the years create an airy décor. Clever
> use of mirrors aid light flow and simple window treatments include handmade
> blinds of French toile with Italian bubbles... Rosie has grown to appreciate
> French country style. "Having spent much time there, I think the French concept
> of country living is beautiful. I have tried to adapt that to Australian life" ...
> The kitchen contains an 1880s meat safe ... while a battered butcher's block
> contrasts with the gleaming stainless steel range (CS June 2011).

> Lush textures, intricate patternwork and design durability herald the latest
> looks in flooring ... Whether reclaimed or re-milled, recycled hardwood is an
> environmentally savvy way to enhance your floor. Imperfections add character
> and a sense of history (CS June 2011).

> There's nothing more fitting than a combustion cooker to take centre stage in
> the kitchen ... There is something of a black art in coaxing alight a vintage
> Aga for the first time ... And celebrated Aga owners include Jamie Oliver, Rick
> Stein, Madonna, Sting and Elle Macpherson – should you ever find yourself
> wondering about that sort of thing (CS June 2011).

> Mix industrial and vintage elements with new to create bedrooms with
> personality (CS June 2011).

Restoring the physical heritage reflects both a degree of Englishness, from
the country homes to the furniture in homes and gardens, and the virtues of

the past. It is a primarily feminine restructuring and regeneration, despite male labour, that espouses a 'return to simplicity', with a focus on the pastoral idyll and romanticism which celebrates the appeal of childlike 'rusticity' over the 'artificiality' of modern convenience (George 2008: 828). Modern conveniences are mere afterthoughts.

Good Food and Wine

The single most dominant theme in *Country Style* is that good food and wine are an intrinsic part of country life and are part of the environment, whether in markets, shops or restaurants. Good food is intrinsic to home cuisine, best consumed on verandas. Food production at every scale takes on heritage qualities.

> Ruby grapefruit and Campari marmalade is just one of the seasonal preserves made by Carol Ruta and Ian Grey. There's plenty more to choose from at South Coast Providores, their lovely shop in Berry (CS June 2011).

> These days when he is baking an Irish *bairin breac* loaf, rich in butter, spice and fruit, or his classic *pain rustique* sourdough, Jason is doing his best to ensure traditional pastry-making and baking skills don't become a "lost art" in Australia (CS June 2011).

Damper however has disappeared. Advertisements, especially for cheese and pastries, develop similar themes. Certain places express the relationship between food and identity particularly well:

> There is a sentiment around South Australia's bountiful Barossa valley that the heart and soul of this place can change a person. You may hear it from the providore who cuts you slices of home-pressed salami, or from the vintner that just poured your second tasting of shiraz, but the odds are good that it will crop up. The words are just as likely to come true in this magical place, which unfolds from the hills 60 kilometres north east of Adelaide. Unlike some other lush well-watered wine regions in Australia, the Barossa valley makes a virtue of colours and contrast – the parched yellow hills are ruled neatly by verdant rows of vines. The roads between vineyards are usually crushed white granite and they glitter beneath cornflower blue skies, dotted here and there by fluffy cumulo-nimbus clouds. If the vineyards are the heart and the wine its lifeblood, then food is Barossa's soul (CS June 2011).

> Orange is a sophisticated food and wine town but when the caffeine siren sounds, educated palates hurry past the sleek main street cafés to a former butcher shop. Try a plunger made with giant Nicaraguan marogype beans and perhaps a chorizo and grilled haloumi roll (CS June 2011).

While feature stories may be relatively subtle, advertisements in such magazines do not seek to be, as in a tourist advertisement for the Limestone Coast of SA:

> I might not be back on Monday. Time moves a little slower... when grazing the menu. The days meld into one another, the wine seems endless and the food actually tastes better. Indulge the senses in the Limestone Coast where world class wine blends brilliantly with fresh local produce (CS August 2009).

Key places, especially those linked to wine regions, recur with particular regularity: notably the Hunter Valley Wine Country and the Barossa Valley.

Country Style has an annual Country Chef of the Year award:

> When Country Style launched its Country Chef series, even the timing was delicious. We spotted a new interest in the food philosophies of regionality and seasonality. Not only in Australia, but around the world, top culinary talents were relocating closer to farmers and growers. Their goal was to produce great food with an integrity, a quality, a sustainability and a regional and seasonal relevance. This just wasn't possible in cities any more. Country is the focus of everything that is exciting in Australia's contemporary foodscape ... to further beat the drum for the philosophy of thinking globally but sourcing locally (CS June 2011).

Food should be produced, marketed and consumed locally. After moving to the country,

> he now turns out several varieties of delicious crusty sourdough bread, alongside a selection of cakes and every three weeks or so a small batch of chocolates. In April the couple opened a rustic French-style store, Van Leuven Culinaire, selling their own and other local products: "We have our own roasted coffee, a selection of teas, salt rubs for lamb and beef made with herbs from our garden, fresh hand-cut pasta, nice lavosh and grissini to go with cheese ... and then there are the jams: raspberry and rosewater, vanilla and marmalade, vanilla syrup (CS June 2011).

> As a pioneer foodie in the region...Rosie loves to entertain, gathering ingredients from nearby farmers' markets, keeping hens, and growing her own herbs and vegetables. "We also have great local produce like olive oil and goats' cheese" she says (CS June 2011).

Food and geography are interconnected: 'we have the Irrewarra bakery nearby, with its wonderful sourdough. There's Otway Pork, Timboon cheese ...' (CS August 2011). Wine is seemingly ubiquitous and is produced in 'wine country'. 'Enjoy a taste of Italy in north-east Victoria, where family-run wineries offer 'romance in a glass' (CS August 2011).

Craftsmen (and women) and manual labour – made with patience and care, not speed, and without (visible) machinery or hired workers, underpin food and wine production, and spill over into other forms of small scale activity. Work rarely constitutes employment, while working alongside wife/husband or kin, pointing to traditional values.

> A new life in rural Tasmania prompted Haidee Lindell to turn vintage fabrics and needlework pieces into stylish children's clothes … while the rest of the family is working at Dad's vineyard (CS July 2011).

The country has gradually become revitalised as a place of creativity, a land of painters, artists and potters, and of studios, craft workshops, wineries and bakeries.

Manual labour is attentive to detail, and therefore small quantities (the antithesis of mass production), and accessible to the consumer at 'cellar doors' and in 'craft shops', where production processes (that eschew mechanisation) are visible. The connections between products, places and people are established, validated and performed. The production of bread is as much an artform as the production of art and pottery.

Consumption

Not only do the stories metaphorically consume the countryside, they simultaneously paint the country as a place for consumption, which requires and sustains a particular ambience. Wine and good food are to be produced, but also to be consumed, preferably in situ, as close to the source as possible. Gastronomy offers solace through local and 'traditional' food cooked, according to inherited recipes (Bessière 1998). Heritage is literally consumed. Consumption is the outcome of production,

> She sits on an enormous velvet couch, crowded with colourful cushions, sipping on freshly picked lemon verbena tea (CS June 2011).

> The ritual of reaching for the kettle is a relaxing way to improve well-being (CS September 2012).

Restaurants have local qualities:

> Magpie's Nest Restaurant is set in old stone stables where you look out over vines and olive trees. The restaurant has a kitchen garden that's always pleasant to walk in and there are plenty of magpies sitting on wheelbarrows and fences, just generally being magpies (CS June 2011).

At a restaurant (The Tea Rooms of Yarck) in north-east Victoria:

'In Sardinia my grandfather had sheep, goats, vineyards, fruit trees, olive groves in the mountains, and grew his own grain. I'd go up the mountain riding a donkey in the morning to help him. It was a very old-fashioned way – and I'm trying to do the same things. I remember the smells and what things should taste like. There was a love to it.' On his farm Pietro runs Boer goats, Dorper sheep and Angus Lowline – cross cattle for meat. He grows all kinds of vegetables, olives and other fruit, and keeps chickens, a few horses and, for nostalgia's sake, donkeys (CS June 2011: 70).

Typically such artisans are referred to familiarly by their Christian names. Even making chutneys relishes and jams is about 'preserving memories' (CS, July 2010).

And in the Barossa Valley:

A more recent international influence is of a Vietnamese flavour and it's being hailed by locals and visitors as a something of a revelation ... Tuoi's dishes ... receive all their greens from the kitchen garden, which reminds her of the way home-grown produce 'used to be' in Vietnam (CS June 2011).

As long as the cuisine is sophisticated, an international presence is more than acceptable: 'For us cooks and chefs in the [Barossa] valley...it's all about cooking with the rhythm of the seasons. Even though we're cooking different styles and flavours we are on the same path' (Maggie Beer, CS June 2011). 'For Italians cooking delicious food for their families is everyday living. And for visitors, it's all very real and authentic' (CS August 2011). Food represents one of the very few places where cosmopolitanism breaks though, Englishness is marginalised and a hint of a multicultural Australia emerges.

Time

The country offers a slower pace of life, less pressure and more time for relaxation and family, tranquillity and time to achieve harmony with the landscape.

I take [my four-year-old] by the hand and walk to the post office – who knows what we may encounter; the beautiful autumn day, a snail crossing our path, colourful plane tree leaves, butterflies and, best of all, the happy chatter of my little treasure. Thank you for reminding me to stop and smell the roses (CS June 2011).

Beauty is all around us – you just have to stop and look... Sadly I think people are losing their ability to understand nature and the simple treasures that are right under their noses (CS July 2011).

The closest thing to a traffic jam is a herd of Holsteins crossing the road at milking time' (CS June 2010).

And for those unfortunate enough to live in cities: 'If I haven't been out there for a while I can feel a sort of tension in myself. Even if it's only for three days I can unwind and recharge my batteries' (CS March 2012). Time is available for the good things in life.

Place, Community and Scale

Relationships with local communities are somewhat muted partly because *Country Style* focuses rather more on privatised space – the home and garden and their immediate environs – but at the very least long-term inhabitants are in the background, if not the foreground, providing effusive welcomes and country solace. Several new residents are not beyond offering homilies on the virtues of engagement and participation:

> Community plays a big part in Rosie's life. 'When I moved here I didn't know anyone but wanted to get involved so I developed my garden for inclusion in the festival'. Over the years Daffodils at Rydal has raised considerable sums for charity. 'It's amazing how such a tiny community can achieve so much', says Rosie who also plays organ in the little church that residents banded together to preserve (CS June 2011).

> They took a risk when they moved to the country but Craig and Roberta Schablon have built a new life where neighbours aren't strangers ... 'In our first week we had everybody driving up to introduce themselves. Within days we knew everyone within five kilometres' (CS July 2011).

> Castlemaine represents the village values that have disappeared elsewhere, and they appreciate being part of a community of creative individuals going about their business with passion ... 'There's a generosity here and a very down to earth community' (CS September 2012).

> I am finding such a close sense of community here. One of our neighbours gives us as many vegetables as we can eat, we get bags of fruit from another friend – and we've even found abalone hanging from our garden gate. Friends invite you to dinner and everyone is cooking something really beautiful from local produce (CS July 2011).

> Donations of stone and firewood and a shed raising where new-found mates 'chipped in with timber and muscle' evoked the spirit of country life (CS August 2011).

Neighbours are not strangers, and provide almost instant food and community, implicitly the antithesis of urban Babylon, but quite unlike the much-vaunted but absent 'mateship' of earlier times (Oxley 1974). Neighbours and others are met in social rather than economic contexts. Rosie's involvement is relatively unusual (and may reflect the fact that she is a single woman rather than the couples and nuclear families who otherwise dominate). Otherwise social interactions and social groups are rarely mentioned. Artists and craftsmen are sometimes solitary folk. Voluntary work and employment are absent. Shopping (beyond boutiques, craft stores and famers markets) is rarely mentioned. Other services, such as education and health, or sports clubs, play no role.

Children and Family

Time and nature provide a safe place for nurturing, close to nature and relatives:

> They chose to settle in Naracoorte, tucked away in the bucolic wine country of south-east Australia… "Naracoorte is where I grew up, and Mum and Dad still live here. There's no violence and it's safe…I'm so glad we came here – in Paris Chloe [our 4 year old daughter] would not have this space or even a backyard to play in' (CS June 2011).

Even global cities thus pale in comparison. Children should and can be close to nature – and to relatives:

> It's wonderful to have our kids growing up with grandparents and older people around …. Most mornings Haidee walks Tom along the waterfront to kindergarten before dropping in for a latté at the restaurant. Then it's time to hit the studio (CS July 2011).

In winter:

> Why not encourage your child to draw some leafless deciduous trees and then hunt for real evergreen leaves to paste into their picture (CS July 2011).

And other seasons:

> [The children] enjoy what each season brings: toasting marshmallows on the bonfires in autumn, running in mud during winter, feeding the piglets in spring and, of course, spending the summer evenings climbing trees (CS August 2011).

Children acquire moral values and are safe and healthy.

While it might be easy to ridicule the caricatures, clichés, adjectives and constant repetition of themes and phrase, *Country Style* is contributing to, partly defining and ultimately performing a new rurality, alongside invocations to 'dream' and 'imagine' (Edensor 2006). That 'new' rurality is retrospective and nostalgic, even what may be seen as some kind of ancient rurality – perhaps where we all came from a few generations ago (Baylina and Berg 2010: 280). The country provides a simpler life in a timeless place where seasons remain dominant and shape lives and aesthetics. In many cases almost every theme identified above is present in single features. Others are implicit and self-evident and require little re-iteration and emphasis. Generic traits are widespread. *Country Style* is far from alone in evoking a particular set of images but shares similar perspectives and sentiments with numerous other glossy magazines, notably *Australian Gourmet Traveller,* and several regional variants. Similar trends exist elsewhere with a slew of glossy magazines in developed countries like Spain, Sweden and Norway (Baylina and Berg 2010; Jonasson 2012) and the United Kingdom, notably *Country Life,* likewise extolling the virtues of the country.

An Elitist Affluent Country

Country Style has constructed a gentrified, elitist countryside: a rural idyll where places have been reconstructed in order to satisfy anticipated, usually distant and virtual, gazes of discerning readers and consumers. Many stories and features have almost identical formats, from the same 'school of creative writing', producing a country and its products incessantly qualified and boosted by a range of adjectives ('delicious') and adverbs ('leisurely'). Fruit and vegetables are always 'freshly picked'; restoration is 'loving'; buildings are 'quaint'. Animals (dogs, horses, sheep) are all particular, often rare, breeds and houses and food have exotic names and even qualities. Consequently 'adjectives like 'traditional', 'authentic,' 'historic' and 'original' are overworked and overdetermined' (Peace 2011: 31) in the quest for retrospection, nostalgia and middle class solace.

Multiple identities, realities, meanings, sites and impressions are conflated and reconfigured to create a sense of place that, for some, provides an attractive and reassuring image that is simultaneously heritage, international and English. The literal restoration of the past, and its more tangential significance in festivals or craft products, emphasises that it is almost literally 'another country'. Places that are both local and global (the foreign phrases and the adverts that offer 'beautiful country homes all over Tuscany') simultaneously recognise that 'the taste of place has become a transnational mode of discernment' (Trubek 2008: 94) and that discernment is to be found in rural Australia.

Markets and craft stores provide a performance space in which are enacted 'exchanges between petit-bourgeois producers and well-heeled middle class consumers …people with more or less the same economic rank and endowed with much the same cultural capital' (Peace 2011: 37). But transactions are not commercial. There are not even throwaway lines about the costs of purchase and transformation of properties, or of exclusive handicrafts, even in advertisements.

That is too vulgar – if you need to know the price you can't afford it (except perhaps of Agas which range from $10,000 to $30,000). Space is being constructed for progressively more affluent users.

The country is not merely physically rural, in the almost complete absence of a built landscape beyond individual buildings, but a place where harmony reigns supreme. The rural is thus the place of preservation of community values, social stability and individual safety – the place myths of a romantic and symbolic countryside (Hopkins 1998) – that are anti-urban, timeless and untroubled.

The romantic construction of the rural involves both stories and pictures, that are 'thick with longing for a near-lost agricultural world and its fading culture' (Peace 2011: 25). These feature residents, many new to the country, refusing to let die and actively restoring, physically and ideologically, homes and ways of life, redolent of other times.

The country is a white world (although there is no hint of 'white flight') where the demographic structure has a remarkable absence of youth (and the elderly) since the dominant market segment are baby-boomers, capitalised and on the cusp of retirement. That is reflected in the photographs of people that do appear, primarily seemingly upper middle-class white women of a certain age, and their husbands. ('Partners' are absent from *Country Style*). Pictures are mainly of objects – houses, gardens and food – rather than people. Those that are there are the proud middle-class, well-dressed, owners of such things – or artisans at work in appropriate 'traditional' garb – supplemented in the advertisements by younger, more svelte models.

'Cosmopolitan' and 'multicultural' are absent words that resonate only in the diversity of cuisine, through 'Memories of Sardinia' and Japanese or Italian cuisine, or at festivals and performances, such as the Bundanoon Bush Dancing Weekend which includes 'Estonian folk dancing to Irish and Scottish reels' (CS June 2011). Both contexts offer alternative versions of nostalgia. Aborigines are conspicuous by their absence, other than at festivals. At the Barunga Sports and Culture Festival (Northern Territory): 'share in the Jawoyn culture, from traditional dance to basket weaving and feasting' and at Laura (North Queensland) 'a two-day festival of Aboriginal culture including dancing, singing and short films' an essentialist Aboriginality is constructed, though no magazine features cover such events. And, despite several features on Daylesford (cf. Gorman-Murray *et al.* 2012), this country too is a heterosexual world.

'Rural gentrification as permanent tourism'

The country emerges from *Country Style* as a place of leisure rather than production, and while many residents are depicted, or depict themselves, as having successfully moved or returned to the country (with 'seamless adaptation'), the country is more likely to be a place for occasional visits or holidays. More obviously the language, phrases and paragraphs used to describe places in *Country Style* are redolent of travel brochures, as the recreational function of country is emphasised. Likewise the presumed greater availability of time offers the possibility of a more recreational lifestyle (Hines 2010). Gastro-idyll and tourist idyll are combined (Bell 2006) in a region of recreation and consumption:

> We have just returned from a fabulous two weeks holidaying in rural Victoria. At times it felt like we were on a *Country Style* tour as I announced to my husband: 'There's a great café/garden/antique shop/farmers' market in this town. I've read about it in *Country Style*' (CS June 2011).

Festivals, shows and markets offer authentic and performative action. *Country Style* favours festivals and markets that are linked to food and heritage. At Margaret River Farmers Market: 'you can meet local producers and growers'; at Bairnsdale 'don't miss this wonderful market featuring many hand made goods' while Wirrabara Producers Market ranges 'from diverse products to plants and woodwork'. At Westbury 'expect freshly picked vegetables, fruit, just-baked cakes, books, plants, food stalls and a friendly vibe … opposite the village green'. The Hunter Valley 'will be buzzing with themed events from high teas and leisurely lunches to cheesemaking classes'. The York Gourmet Food and Wine Festival offers a 'showcase of food and wine from all over WA, along with fresh produce from the Avon Valley, including olive oil, chilli, yabbies and exotic spices'.

> Find fresh produce and handmade wares at a country market, step back in time at a historical festival, or simply sit back and enjoy one of the many music events (CS June 2011).

Festivals and shows are a regular feature. Cooktown Discovery Festival 're-enacts Captain James Cook's emergency landing on the banks of the Endeavour River in 1770'. At its Village Alive Day the 'historical village [a word rarely used in Australia] of Loxton honours the 19th Century pioneers who battled drought, flood and rabbit plagues in the district. Period costumes are encouraged'. Music, notably jazz, are more fleeting accompaniments to food and wine. At the Tamborine Mountain Scarecrow Festival 'drive along the 14-kilometre Scarecrow Trail, and enjoy the local cafes and pubs'. *Country Style* thus offers a 'kind of release, a fleeting freedom …from the late capitalist mainstream of supermarkets, expressways, office towers and city neighbourhoods' (Peace 2011: 31) – nostalgia and drive-through heritage.

Beyond festivals the country is a place for passive not active recreation, observing rather than engaging – that may reflect both a middle-age and female readership – although there is little space for art galleries or museums. Sport and even exercise are absent, even when potentially individualist (bushwalking, fishing, cycling, golf, riding). Furnishings, food and fashion, rather than fox-hunting, reflect a certain essentialist female lifestyle. Recreational activities (beyond restoration) include cooking, reading, painting and sewing: broadly female recreations. Men have no recreations in *Country Style*. Football, shooting and vintage cars are inconceivable.

Post-productivist Countryside

Country Style offers no hint of countryside in crisis, whether social, economic or environmental, and no mention of politics or economics. All is seemingly well in the soft countryside of consumption without cost. Economic growth happens elsewhere. Employment is never mentioned other than in some forms of self-employment. Work is only undertaken by hand in the craft production of food, wine, pottery and similar goods. Such workers are 'historically aware and singularly committed small-scale entrepreneurs' (Peace 2011: 24) and skilled tradespeople. Production is often undertaken by families, husbands, wives and children together, occasionally with inherited skills: 'the petit bourgeois family with its close and enduring relation to land' (Peace 2011: 26). Work is intensely pleasurable and creative, and valuable in retaining and restoring past skills and knowledge, invoking a time when agriculture and other forms of production were trade rather than industry, and food was natural rather than synthetic. Work needs patience, care and individualism, not speed, machinery and homogeneity. Not only food and wine but pottery, art, handmade clothes and other artisanal produce, literally even a 'patchwork economy', are marked by the cultural distinction that comes from tradition, knowledge and care – infused with the past.

Concern with contemporary agriculture barely extends beyond niche activities involving interesting and unusual plants and animals. Mining unsurprisingly plays no role. Male work, other than in food and wine, is largely absent from this somewhat gendered countryside. Female work is centred on kitchens and home gardens, focused on caring, nurturing and creativity.

Only implicit in accounts of markets, food and wine, is the possibility that goods are bought and sold. But 'consumption has become a serious form of work …the discipline of learning to link fantasy and nostalgia to the desire for new bundles of commodities' (Appadurai 1996: 82). True artisans have created what amounts to a 'new bundle of commodities' and the discourses emanating from *Country Style* effectively offer them 'to middle class consumers through a metaphorical prism of nostalgia and escape, longing and desire, fantasy and dreaming' (Peace 2011: 25).

The most fortunate households are those who have been able to move (back) to the country and realise their dreams. Just as in similar glossy magazines in Spain and Norway 'they never regret having migrated to rural areas and confirm the vision of the good life in the countryside' (Baylina and Berg 2011: 288). Their own adjectives perfectly match those of the feature writers. Their return is wholly independent of attempts to market the countryside in the absence of any focus, even mention, of service availability, economic growth, employment or affordable housing. Indeed the new residents in *Country Style* are akin to those of British residents in south-west France where 'rural France is presented as the rural idyll, characterised as the Britain of 50 years ago, offering a way of living that the migrants believe is no longer available to them back in Britain' (Benson 2009: 123). The past becomes another country.

Just as economic activity is imbued with culture rather than economy *Country Style* is nostalgic rather than modern. While it is reasonable to assume that both readers of *Country Style* and those featured in it have some access to cars, mobile phones, computer technology and supermarkets, other than an occasional old-fashioned (circular dial) telephone, they are utterly invisible. Indeed the pictures rather than the text are most evocative of an earlier era. Carefully selected and restored objects display tasteful consumption. Modernity mainly intrudes in advertisements where upmarket kitchen utensils and other forms of technology appear.

Services are non-existent. While marketing rural places centres on service provision, so that, for food alone, small towns like Glen Innes triumph the size of supermarkets and the arrival of McDonald's, in *Country Style* its arrival would be the kiss of death and utterly ignored (Connell and McManus 2011, Peace 2011) as seriously threatening the proper primacy of slow gourmet food and wine. Likewise while marketing stresses good affordable housing, *Country Style* has no mention of motels, semi-detached houses in streets, let alone caravan parks, or the towns where they exist. People's everyday lives are absent, since that might necessitate addressing practical and contentious issues such as the availability of services (education, health and transport) and more general rural policies (even the role of agriculture as production not aesthetics). What that would also do is 'clearly locate the representations in space and time' (Baylina and Berg 2011: 287), something that is avoided in the timelessness of *Country Style*.

A Distinct Geography

While generic themes typify *Country Style,* and similar magazines, and there is a strong implication that the actual place is unimportant, that idyllic country is ubiquitous and the rural idyll placeless, they nonetheless do occur in particular places. Many such places would be unknown to most readers, and are never identified on a map, and rarely by reference to a nearby town, rather than being in a valley, across a creek or amidst hills: lifestyle not specific place is important. Nonetheless such places are centred around some of the 'hotspots' of rural gentrification, such as Berry and the Barossa Valley, Mudgee and the Hunter Valley, the Margaret River, Gippsland and the Blue Mountains. They make up a generic countryside for the discerning retrospective gaze.

These hotspots are close to coasts, rivers or lakes, within a rolling countryside where views and vistas are possible, and away from the dry, 'boundless plains' further inland. Almost every featured place is within 200 kilometres of state capitals, and usually rather closer. Such places are accessible to a metropolitan market. Quirindi, in the Upper Hunter valley, marks the inland limit of *Country Style* features, although some parts of Western Australia are also far from Perth. *Country Style* does not venture into the bush; the landscape is thoroughly tamed (and perhaps therefore free from threats of fire). Rather it invokes a number of particular places, like the Barossa Valley and the Hunter Valley, within a day trip from a metropolitan capital, allowing temporary or permanent escape.

The country is also rural. While towns like Berry and Ballarat are mentioned in features, they serve simply to locate rural residents, who usually live outside them in isolated homes. Villages are mentioned more frequently despite the concept of village being both alien and English in landscapes of dispersed settlement. Indeed even the existence of towns, especially large towns, is disguised. A feature on a restaurant on the New South Wales South Coast consistently referred to the coast until the final line which revealed that is was located in the industrial town of Wollongong (CS June 2011). Geography is invented, sanitised and anaesthetised.

An Urban Idyll

Country Style has created and nurtured a distinct vision of country Australia: a rolling, green aesthetic landscape of restored houses and crafts, good food and wine, community and morality, and values retained or restored from a pre-industrial past. Selective components of heritage, tradition and authenticity demarcate the country. Tradition and geography are simultaneously invented. The country is tied by an umbilical cord of access to the coast and only exceptionally extends beyond a 'sandstone curtain', somewhere around 100 kilometres inland, that separates the coast from the inland.

There are other rural and regional Australias beyond and within that line, and multiple post-productivist and productivist countrysides, few of which resemble the rural Australia of *Country Style*. That is so evidently true from its sub-title: 'Escape the Everyday' (Figure 11.1). That 'everyday' is almost certainly urban rather than rural. Indeed it is much easier to find *Country Style* in the eastern suburbs of Sydney than in, say, the newsagents of small towns like Parkes. But it might equally be a rural 'everyday'. *Country Style* has almost nothing in common with media such as The Land, other than a loose focus on the 'rural', and even less in common with the readership of other monthly glossy magazines that can easily be found in Parkes – such as *Boar it up Ya* ('the only hunting magazine with grunt'), *Aussie Boar Hunters, Bowhunting, Bacon Busters* and *Australian and New Zealand Handgun* – all of which offer distinct perspectives, a quite different gaze upon, and possibilities for leisure within, rural and regional Australia. Unlike *Country Style* they cannot easily be found in the eastern suburbs of Sydney. Vast areas of rural and regional Australia are absent from the pages and vision of *Country Style* (and, indeed, from *The Land* or *Boar it up Ya*). Likewise *Country Style* is far from *Wake in Fright*, the 1971 film of an urban teacher awash in the tough outback mining town of Bundanyabba, and whose cinematic legacy lingers on. There are other inland Australias.

Country Style provides a selective invocation of the past – a nostalgic return to an invented timeless tradition (harmony, crafts, community, family) – the absence of colour, bumptious youth or the working class, and the vicarious marketing of what is often no more than outer metropolitan Australia. It is Australian but international. This country is anti-modern and the antithesis of what regional planners seek to promote: a post-productivist (or perhaps pre-productivist) elite, themed and designer countryside of consumption rather than production – an idyllic space for jaded urban residents, with leisure time and cash in search of cultural capital: social stability, suburbia with soul and modernity on hold, where land, food and families are nurtured, which is intrinsically virtuous. Dissident voices are absent. Rurality is safety, purity, middle-class prosperity, family, community and place of leisure, of femininity rather than masculinity. Indeed it is curiously post-rural with the country merely backdrop. In the end *Country Style* (published in the inner city Sydney suburb of Alexandria) tells us much more about urban life – and the sandstone curtain – than about contemporary rural life, and even more about the country of imaginations and aspirational dream time.

Acknowledgement

I am grateful to *Country Style* for permission to publish the cover image and the various quotations.

References

Appadurai, A., 1996, *Modernity at Large: Cultural Dimensions of Globalization*, Minneapolis: University of Michigan Press.

Baylina, M. and Berg, N., 2010, 'Selling the Countryside: Representations of Rurality in Norway and Spain', *European Urban and Regional Studies*, 17: 277-292.

Bell, D., 2006, 'Variations on the Rural Idyll', in P. Cloke, T. Marsden and P. Mooney (eds), *Handbook of Rural Studies*, London: Sage, 149-160.

Benson, M., 2009, 'A Desire for Difference: British Lifestyle Migration to Southwest France', in M. Benson and K. O'Reilly (eds), *Lifestyle Migration: Expectations, Aspirations and Experiences*, Farnham: Ashgate Publishing, 121-135.

Bessière, J., 1998, 'Local Development and Heritage: Traditional Food and Cuisine as Tourist Attractions in Rural Areas', *Sociologia Ruralis*, 38: 21-34.

Connell, J. and McManus, P., 2011, *Rural Revival? Place Marketing, Tree Change and Regional Migration in Australia*, Farnham: Ashgate Publishing.

Edensor, T., 2006, 'Performing Rurality', in P. Cloke, T. Marsden and P. Mooney (eds), *Handbook of Rural Studies*, London: Sage, 484-495.

George, J., 2008, 'Small Town Essentials: Constructing Old World Charm in "Backwater" Communities', *Continuum: Journal of Media and Cultural Studies*, 22: 827-838.

Gorman-Murray, A., Waitt, G. and Gibson, C., 2012, 'Chilling Out in "Cosmopolitan Country": Urban/Rural Hybridity and the Construction of Daylesford as a "Lesbian and Gay Rural Idyll"', *Journal of Rural Studies*, 28: 69-79.

Hines, J., 2010, 'Rural Gentrification as Permanent Tourism: The Creation of the "New" West Archipelago as Postindustrial Cultural Space', *Environment and Planning D. Society and Space*, 28: 509-525.

Hopkins, J., 1998, 'Signs of the Post-rural: Marketing Myths of a Symbolic Countryside', *Geografiska Annaler B*, 80: 65-81.

Jonasson, M., 2012, 'Co-Producing and Co-Performing Attractive Rural Living in Popular Media', *Rural Society*, 22: 17-30.

Oxley, H., 1974, *Mateship in Local Organization*, Brisbane: University of Queensland Press.

Peace, A., 2011, 'Barossa Dreaming: Imagining Place and Constituting Cuisine in Contemporary Australia', *Anthropological Forum*, 21: 23-42.

Poole, S., 2012, *You Aren't What You Eat*, London: Union Books.

Trubek, A., 2008, *The Taste of Place*, Berkeley: University of California Press.

Chapter 12

'Not just drought.' Drought, Rural Change and More: Perspectives from Rural Farming Communities

Louise E. Askew, Meg Sherval and Pauline McGuirk

The 'Big Dry', a prolonged dry period in Australia from 1997 to 2009, seared much of the Murray-Darling Basin region and resulted in large agricultural losses, degraded river systems and increased uncertainty in rural communities although climate change in the form of drought is not new to rural Australia (Wei *et al.* 2012). For many years, generations of Australian farmers and farming communities have battled such climatic extremes. However, the most recent drought event competed with a myriad of changes to their lives and as such, the façade of stoicism has slowly begun to crack. This chapter examines the changes exacerbated by drought occurring in rural Victoria and considers the challenges facing both rural towns and farming families, whose economic future and social well-being are predominantly associated with agriculture. By drawing on locally situated knowledge from case studies of the rural towns of Mildura and Donald, this chapter shows how issues such as reduced water supply, increasing agricultural costs, farm succession and cumulative uncertainty are affecting the ongoing viability of people living off the land in these drought-affected rural areas.

Like many other rural towns in Australia where agriculture is a mainstay, Mildura and Donald are experiencing a combination of strains on their communities, townships, farms, and farming families. These pressures arise not only from drought but also from extensive changes to local communities and farming enterprises that include: a rapidly evolving water market, the increasing competition of commodity markets, wide-ranging rural demographic shifts, and changing rural service provision and investment. Drought and the effects of long-term drying of these agricultural regions represent just one challenge amongst a melee of change. In the oft-repeated words of residents from these rural communities, the problems they are confronting are 'not just drought', they are a combination of 'drought and more' which make successful adaptation all the more difficult, particularly when current policy regimes remain inadequate and local experiences little understood (Sherval and Askew 2012).

This chapter seeks to extend our understanding of the issues facing both these drought-sensitive regions and those like them throughout Australia today by exploring the diverse, changing and sometimes strained contexts of rural towns and communities. It suggests that any future provision of support to communities throughout ongoing and future changes will require a holistic approach, rather than one that visualises drought as a once off, crisis-ridden event as government support schemes traditionally have done. Overall, this chapter seeks to develop the discussion surrounding drought impacts, and their embeddedness within a myriad of other rural changes and challenges, by drawing on locally situated knowledge to inform future decision-making nationally in this evolving, yet vital arena.

Drought and More: Decline and Change in Rural Agricultural Areas

Rural towns in Australia are experiencing a diverse set of challenges arising from changing climates, agricultural industries and rural demographics. Climatically, rural towns in south-eastern Australia have been experiencing the effects of a drought, or the 'Big Dry', since the mid-1990s (Verdon-Kidd and Kiem 2009). While the effects of the drought have been dire, equally troubling has been the unforseen reversal of fortunes. This reversal came with the beginning of a La Niña cycle which saw the region recording its wettest year on record in the second half of 2010 and early 2011, ending a long sequence of 15 years without a wet month (Wei *et al.* 2012). Although this led to a rapid refilling of water storages along the Murray-Darling Basin and elsewhere, it did not remove many of the complex issues surrounding water and its allocation, which continue to plague locals (see Chapter 13). Despite this rainfall, the Basin region will continue to experience the impacts of the 'Big Dry' for some time, particularly due to the widespread dominance of agricultural-based economies there. Indeed, the prolonged nature of the 'Big Dry' has had severe impacts on many forms of agricultural production due to the associated declines in water availability and supply. As a consequence, the drought has affected the economic viability of many farms and farming businesses. Estimates by the Australian Bureau of Agricultural and Resource Economics (ABARE) state that the recent drought resulted in a decrease in average farm incomes of $29,000 from 2005 to 2007, and an increase in negative cash farm income (where expenses exceed income) from 24 percent to 42 percent over the same period (ABARE 2008). The ongoing declines in water availability and supply projected for the future, particularly when the El Niño cycle returns, mean that these drought-affected farmers face considerable and ongoing challenges to their agricultural base and economic well-being (CSIRO-BoM 2007, Gunasekera *et al.* 2007).

In addition, rural towns have been confronted with rapid changes to the agricultural and farming sectors, which can represent the primary industry base for these areas. These changes are characterised by increases in farm size, declines in the traditional family farm, and increasing competition and global market pressures. While the contribution of the agricultural sector to the GDP has decreased significantly over the last fifty years, agricultural output has almost doubled since its 1974-75 level (ABS 2009a). To achieve this growth in productivity, the size and technological advancement of farms continues to increase rapidly, with an associated decrease in actual farm numbers and farm employment (ABS 2009b) – a trend Barr (2005) describes as 'get big or get out' (see Chapter 8).

These shifts in the farming sector have also coincided with, and contributed to, significant social and demographic change in some areas of rural Australia (McGuirk and Argent 2011). Overall, the population of rural farming regions is declining rapidly (Barr 2005, BRS 2008). Moreover, these rural communities are also increasingly characterised by ageing populations, decreasing family incomes, low educational attainment and opportunities, and a declining workforce and skills-base (Wheeler *et al.* 2012). Such socio-demographic shifts have been directly linked with decreases in economic viability in some rural towns, indicated by bank closures (Argent 2005), restructuring of the retail sector (Argent *et al.* 2007), job losses and shifting expenditure to key rural centres (Baum *et al.* 2005). This is often exacerbated by the changing nature of rural service provision – under the rubric of increasing self-reliance – which has severely impacted the provision of rural services, health and educational facilities (BRS 2008, Davies 2009). Indeed, Alston and Kent (2004) suggest that many human services in these rural areas were in a state of overload and crisis prior to the recent drought, which has only served to exacerbate ongoing demand for services.

These combined processes can portray a common story of rural decline – one that often oversimplifies the complexities and variations in rural Australia's experiences of recent socio-economic and environmental change. For example, some rural towns have become less economically 'coupled' or reliant upon agricultural markets (Wilson 2009), therefore evading many of the associated impacts of neoliberalism and the globalisation of agricultural trade and production such as the decline of the family farm (Pritchard *et al.* 2007). Diverse rural experiences – such as developments of new agriculture and mining opportunities, increasing rural amenity of some rural 'tree change' towns, and incidences of local entrepreneurialism and cooperative regional development initiatives have enabled any seemingly inevitable trajectory of rural decline to be avoided (see Cocklin and Dibden 2005, Argent and Walmsley 2008).

To avoid oversimplified stories of decline (Botterill 2009), these diverse, changing and sometimes strained contexts of rural towns need to be further explored. This approach is particularly important in the face of the changes posed by future drought events and ongoing climate change. Indeed, the exact relationship between drought, farming and other forms of rural change is often difficult to disentangle; yet it is clear that the experiences of farming communities can be strengthened or exacerbated by drought events. Thus it is in these drought-sensitive locations where insights into what to change, what to cease and how to think and learn about drought are most likely to be developed (Golding and Campbell 2009).

Uncovering 'drought and more' in Rural Farming Towns

The choice of Mildura and Donald as the focus of this chapter is due to their unenviable position of being at the forefront of drought in Victoria. Mildura, a rural region of approximately 58,000 people, located at the intersection of Victoria, New South Wales and South Australia (also known as Sunraysia), has a long history of agriculture and irrigation, which are closely linked to the region's economic fortunes and decline. Likewise, Donald, a small rural settlement of 1700 people in north-western Victoria, has an economy based on dryland agriculture, particularly grain (wheat, oats and barley) and some sheep grazing (see Figure 12.1). Despite their proximity, Mildura and Donald have differing rainfall and climatic patterns, water supplies and policies, agricultural industries, economies and socio-demographic profiles. Thus they have related but different experiences of the impacts of the long-running 'Big Dry'. Together they provide a useful comparison as to how particular communities have experienced extended drought conditions, and how these experiences are part of a broader context of extensive socio-economic adjustments being made throughout rural Australia as a whole.

The discussion is based on in-depth interviews in Mildura and Donald, drawn from representatives of relevant local and regional organisations, government agencies, local councils, private business, and from community groups and farming enterprises. Due to the affiliated and interconnected nature of many of the recruited organisations, some participants represented more than one sector,

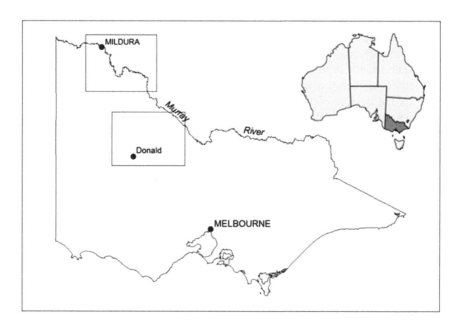

Figure 12.1   Map of Case Study Areas – Mildura and Donald, Victoria

location or organisation. Interviews focused on common perceptions of drought, how farmers and others in the farming community personally experience drought, and what mechanisms they used, if any, to cope with drought, and broader rural change in these locations. By allowing people to talk about their experiences of drought – a process that often evoked very emotive, heart-felt and personal stories, powerful insights were provided into the ways people living in rural locations experience and manage drought as part of changing rural contexts. The following sections of the chapter address Mildura and Donald in turn and seek to provide a contemporary perspective on the rapidly evolving socio-economic implications of drought, water rights and changing farming practices, that are beginning to preface new ways of approaching social, cultural and economic transitions.

'It's not just drought': Mildura, Agricultural Change and Decreasing Water Security

The Mildura region encompasses the city of Mildura itself, together with the key townships of Ouyen, Merbein, Red Cliffs, Robinvale, and Walpeup. Generally, Mildura's social profile reflects that of many rural townships with an ageing population, declining workforce, and lower than average incomes and employment levels (MDC 2009). Yet, Mildura is also positioned as a regional centre at the intersection of three states, which results in the in-migration of residents and workers. This characteristic of the region has resulted in relatively stable population growth and an increasingly diversifying economy, although with a long history of reliance on agriculture and irrigation.

Agricultural production in the region began with sheep runs as early as the 1840s, and by 1886, the Chaffey brothers set up the first private irrigation settlement in the region on the banks of the Murray River. Further state-owned irrigation settlements followed, together with substantial post-WWI and WWII soldier settlements in Red Cliffs and Robinvale. Currently, the Mildura economy remains centred on agriculture, which accounts for 17 percent of the Gross Regional Product (GRP) and 15 percent of total employment in the region (2007-08) (MDC 2009). The main agricultural crops of the region have been part of the landscape since early agricultural and irrigation expansion – including wine grapes, dried vine fruits, table grapes, and citrus – albeit with varying scope and success.

Recent political, economic and climatic conditions, however, have presented the region with some complex problems. The recent drought posed a challenge that confounded an already strained rural setting. The Mildura region had been addressing a rapidly evolving irrigation system as a result of ongoing drying, decreased water allocations and expanding water trading. This change came at a time of record low commodity prices for some of the main regional agricultural products, namely wine grapes, which suffered a severe downturn due to global oversupply and market competition (DPI 2010, Winebiz 2010). In addition, producers of the region confronted a series of fundamental changes to the farming sector including the expansion of farms and farm trade – the 'get big or get out' trend, declines in family farm succession, and increasing uncertainty around crop selection and investment. Some of these issues had been a direct result of drought and a drying environment; however, others were interconnected with ongoing changes to government policy, international trade and agricultural markets. The complexity of the situation is captured in the views of government and industry representatives:

> What we've all been saying, and this is a consistent message ... it's not just drought that has impacted this region. The farming community in this region have had a number of factors come at the same time, and that's never happened before. So you might have had a drought but you had reasonable commodity prices. Whereas now, we've had drought ... there's been wine grape glut, and on top of that, there's generational change in the farming community. I feel it's very much a pivotal point in terms of where we go to from here in farming. (Senior Planner 1, Department of Planning and Community Development)

> It's not just drought. You've got to understand it in the context of all the other pressures. [Farmers] could've coped with the drought, it would have been a lot of stress, but with the wine grape prices, then you can't afford to buy water to keep going. It's an unresolved, slow disaster. (Senior Social Researcher, Department of Primary Industries)

These statements make clear the need to understand the situation in Mildura as complex and 'more-than-drought' and to broaden the analysis of drought in the region, to capture the complexity of and interconnections between the issues facing the region. We consider three key regional challenges in turn: the impacts and experience of (i) water trading, allocations and supply; (ii) falling commodity prices; and (iii) broader changes in the rural and farming context.

The region has been confronted with rapidly changing water security and supply over a number of years. Water reforms, initially introduced under the National Competition Policy in 1994, have continued the process of unbundling water from the land, to create a 'free' water market of tradeable and saleable water separated from the land (Quiggin 2008). The responses to this marketisation process, however, have been varied and range from confusion and outright resentment through to experimentation and learning to manipulate the water market. The following statements capture some of these diverse experiences of the expanding water market:

> There wasn't the understanding of how you manage, all of a sudden, the security of water being threatened. We might be able to manage drought in some respects, it's been the policy issues and intervention in the [water] market which causes a whole range of other issues. (CEO, Mildura Development Corporation)

> [The water market] presents another set of rules and it's getting quite complex. Some operators can really fine-tune the risk in line with their business. Others get very confused by it and caught out by the rules, which is understandable, it's a rapidly changing and complex field. (Consultant, RM Consulting Group)

Engaging with this evolving water market was found, at best, to be 'tricky' (CEO, Mildura Development Corporation) and, at worst, a 'nightmare' (Farmer, Mildura Region) for farmers. A number of producers have lost significant amounts of money in the water trade due to inconsistent variations in market price and water allocations. Indeed many producers continue to express considerable uncertainty in learning how to navigate a shifting and evolving water market, and have had to rapidly adjust their agricultural planning and mindset: weighing up potential water losses, the cost of water versus the value of crops, and the declining value of land unaccompanied by water – a change which has seen many farmers exit the industry entirely.

In association with an expanding water market, farmers in the region are also confronting a variable water allocation system. Seasonal irrigation allocations from the region's rivers, including the Murray, are regulated by state authorities to respond to changing water supplies. Sudden, variable and unanticipated declines in water allocations have characterised recent years. The variability and ad hoc nature of allocations – between NSW and Victoria, from season to season, from forecasts to actual allocations – have induced great local frustration:

> We went from 100 percent [water allocation] to zero. So originally it was guaranteed that 96 years out of 100 there would be 100 percent water allocation, with a worst-case scenario of 60 percent ... and then suddenly, we went to zero which was huge in terms of how irrigators formulated their business. (Senior Planner 1, Department of Planning and Community Development)

> The problem is that you can't bank on how much water you get at the start of the season, which is fairly important for permanent plantings. So at the start of the season we were looking at a grim forecast of 23 to 26 percent maximum, so a lot of people bought water, and now we are sitting at 77 percent. So a lot of people that bought water didn't need to. (Project Officer, Department of Primary Industries and Farmer)

Uncertainty surrounding the Murray-Darling Basin Authority's recently released Draft Basin Plan (MDBA 2011), which suggests further reductions to irrigation supplies in an effort to preserve the river water quality and flow, has heightened tensions further as communities inland and down-stream seek to re-establish a secure future. While the La Niña event of late 2010/2011 proved to be 'a punctuation mark in a prolonged drying trend', there is no doubt that on-going access to water is essential for continued agricultural pursuits into a drier future (Wei *et al.* 2012: 914).

Thus water security is essential for producers to be able to plan, invest in, and irrigate their crops. Many farmers have invested considerable sums of money in new irrigation systems and on-farm technologies to cope with changes to the farming industry, yet are struggling to negotiate the uncertainties of water supply. Government, agency and farming representatives were united in the view that a stable and secure water allocation and buy-back system which could be more readily and effectively negotiated, planned for and managed by farmers was fundamental to reducing the complexities rural producers must negotiate.

The difficulties of drought, exacerbated by inconsistent water governance, are further intensified by a second challenge: the particular impacts of declining commodity prices on the Mildura region. Indeed, the rapidly changing water context comes at a time of collapse in the wine and wine grape industries driven by oversupply in the global marketplace, the rising value of the Australian dollar and the absence of tariff protection. Moreover, about a quarter of the wine grape crop in the region is the now 'unpopular' Chardonnay variety that, in an already collapsed market, represents one of the hardest hit (Winebiz 2010).

The dominance of wine grape crops in the Mildura region overall means the region has increasingly experienced low farm incomes, unmanageable debts, and an escalating number of grape growers leaving the industry (RFCS 2008). Indeed, the areas of wine grape crops in the region that were deliberately left without irrigation reached a high of 26 percent in 2008-09, with local reports of large amounts of crops also left unpicked (Mallee Catchment Management Authority 2009; see Figure 12.1). A decision by growers not to irrigate crops is significant and reflects the enormity of financial stress incurred by a changing market and agricultural context, as described here:

> The traditional 20 to 30 acre fruit grower is having to compete with international markets. And then with the price of water … I've got friends who are just not watering their vines, they've said that they're not picking, it's not worth it, they're not covering costs. (Manager, Mallee Family Care)

> It's sad because we're losing good growers still due to commodity prices. There's a lot that have done everything they can in terms of water saving and still can't survive … they just haven't got the reserves. So people have learnt how to manage with less water to a degree, but the low commodity prices were the killer. It's sad to see, you drive through here and there are so many dead vines. (Project Officer, Department of Primary Industries and Farmer)

Regional agricultural reports indicate that other producers have suffered commodity price fluctuations, such as citrus and dried fruits, whilst other sectors are flourishing, including table grapes and sheep producers (MDC 2009, DPI 2010). However, with the dominance of wine grape producers in the region, the severe downturn in the industry has had resounding effects on the productive capabilities of the region, as more leave the land and farming entirely. The combined pressures

Figure 12.2 Grapevines Left without Irrigation in the Mildura Region, Victoria

of drought, water trading/allocations, and commodity prices has meant that, unlike in previous downturns, few producers have the financial reserves to shift to other industries as they were able to in the past. The result was a widespread uptake of a Federal Government 'small block irrigators exit grant' by wine grape producers, leaving blocks of once productive land stripped bare and contributing to the sense of stress felt throughout the regional community (see Figure 12.2). The main federal exit grant received much criticism due to the provision that all permanent plantings, irrigation systems and infrastructure be removed from the land, with no irrigated farming activity to be undertaken for a period of five years after exit – essentially leaving strips of bare land throughout the region.

A final challenge that further complicates the more-than-drought conditions facing the Mildura region is long term shifts in the farming sector more broadly. Some of these are related to issues of drought and commodity prices, while others are part of the sectors' response to long-running changes to global markets and rural demography. These, in combination, present unprecedented scenarios that infer confronting conditions threatening the sustainability of rural communities, social networks and economies. The effect of the ongoing rise of the large amalgamated farm and multinational agri-business was registered by Mildura representatives in the following ways:

> Smaller growers have limited capital and limited opportunity for change. They're the mum and dad partnerships ... the children go off to tertiary education, good jobs in the city, and they don't want mum and dad's farm. Mum and dad are sitting there with their labour force gone ... some just haven't got the finances

to change anything. And there's so much doubt in the industry that they're not game to. The family farm has just about had it. (Coordinator, Rural Financial Counselling Services)

They're not passing the farms on now, they are staying on longer and longer ... that succession stuff is not happening as it was. The neighbours are not buying you out which used to happen ... it's now the large enterprises. (CEO, Mildura Development Corporation)

Farmers in the region are experiencing a combination of industry and demographic pressures that have been in process for decades. While in some regions, there is evidence of the uncoupling of rural towns from a reliance on agricultural production (Wilson 2009), in areas such as Mildura, agricultural ties remain strong and thus changes to agricultural production conditions challenge traditional farming families and the broader community.

The complex accumulation of environmental, economic and social changes presents a series of challenges for Mildura farmers. These are sometimes experienced as recent and compounding phenomena for irrigation farmers – presenting a 'huge learning experience, in which there's no living memory to draw on' (CEO, Mildura Development Corporation). Moreover, in an irrigation area such as Mildura, these experiences are also driving changes to the attitudes and traditions of irrigation in the region. Irrigation settlements are privatising, subdividing and modernising infrastructure to maintain efficiency, and producers are increasingly opting to exit their landholdings, or shut down entire irrigation districts (such as the Campaspe irrigation district south-east of Mildura). Yet, the traditions and beliefs in irrigation supply are engrained in the Mildura region. As noted in the Productivity Commission Inquiry into drought support (2009: xx) 'irrigation drought is uncharted territory' and with this comes change, uncertainty and fear for irrigation farmers. For some farmers, the uncertainty has been too much, and they have left farming entirely. For others, they are learning and creating the practical knowledge base through which to adapt to living and working with less water and as part of a changing farming landscape. Across the experiences described here, there is a sense of ongoing change, doubt and a restless tension as people weigh up their future – presenting significant questions around the impacts such uncertainty has on the economic and social well-being of the region as a whole.

'It's more reliable than drought': Rural Change and Farming in Donald

Unlike the irrigation districts of Mildura, the local economy of Donald is based on dryland agriculture (see Figure 12.3). The agricultural economy has historically survived without major sources of water except rainfall. The recent introduction of the 'Wimmera-Mallee pipeline' though, a joint project of the Victorian and Federal

Figure 12.3   The Dryland Farming Region of Donald, Victoria

Governments and Grampians-Wimmera-Mallee Water Corporation, has resulted in some improvements in water conservation and supply to the area (Victorian Government 2010). Despite such improvements, the agricultural-based economy of the region has suffered declines over recent decades due to a combination of challenges and changes that are addressed in turn below: (i) climate change; (ii) shifts in agricultural industries and markets; and (iii) rural socio-demographic contexts.

The agricultural producers of the Donald region rely almost totally on rainfall, therefore the critical issue during the recent drought has been a lack of rain, but more specifically the failure of autumn rains, a lack of late spring rainfall, and the coincidence of hot weather and extreme rainfall events (Verdon-Kidd and Kiem 2009). Indeed, grain farmers in the region had suffered a succession of 'bad' years characterised by reduced yields. The 2009 season in particular was associated with a very poor crop due to below average rainfall throughout late 2008 and early 2009, followed by extreme hot weather and heavy, crop-damaging rainfall. Recurrent below average years created considerable issues of debt and stress for farming and local communities in the region (Birchip Cropping Group 2008). Increasing uncertainty about when and where rain will fall, along with other weather conditions (e.g. heatwave, strong winds) depleted farming communities' emotional and social resources.

The local impacts of declining rainfall have also been overlapped with a period of successive price reductions for grain globally. Unlike Mildura, the problems of market pricing are not so much the result of commodity gluts, but rather the competition from good quality harvests in the northern hemisphere and a low effective price due to a rising Australian dollar (DPI 2010). Many grain farmers felt that this change, combined with generally fluctuating grain prices, increased the difficulties faced by farming businesses in the region, as illustrated here:

> A lot of our problems at the moment are prices ... so at the moment, on top of drought, is the market driven stuff. And they're saying the prices aren't going to get any better. (Rural Services Officer, Centrelink)

> Some [farmers] are just keeping their head above water. The unfortunate bit is that when we get good yields here, prices drop. And if we get poor yields the prices rise ... but if it gets too high, they import grain ... that's just lunacy! (Dryland Farmer, Donald)

> The effect [of commodity prices] on the farm has been astronomical. They are not getting paid ... [The supermarkets] just won't pay ... The farmers have got to give it to them for next to nothing. (Coordinator, Donald Community Centre)

Unlike Mildura, in the dryland areas of Donald, the problems of commodity prices are not so much the result of a commodity glut but rather, in the cropping sector, farmers having to adjust to selling grain without the 'single desk' of the Australian Wheat Board (AWB). The single desk represented a central body through which to market grain globally. It gave growers the market power to achieve supply chain efficiencies and reduce costs. Under recent wheat market reforms, the single desk was replaced by a free market system of marketing and exporting grain. In the face of already declining commodity prices, this shift has added pressure on farmers as they are now responsible for marketing and selling the grain as well as growing it. As the Mayor of Buloke Shire Council (Donald) explains: 'the farmer now has to market [grain] himself ... so he is harvesting, stripping, carting, trying to sell the stuff, and find the right price ... it's just so difficult'.

In addition to declining commodity prices, as in Mildura, farmers in Donald confront broad ongoing changes to the farming sector. Cropping farms are growing bigger to compete in global markets, and are becoming more technologically advanced to gain efficiencies in on-farm practices. The flow-on effects to the composition of farms and small farming communities such as Donald are immense. The fine interconnections between these effects are captured in these insights from locally knowledgeable government officials:

> Every thirty years in the cropping area ... the number of people you need to run the farm halves, and the farms double in size. And because they've got bigger machines and more complex machinery, the skills to keep it going and

service it are much different to what they used to be. So those traditional services withdraw out of towns like Donald. (Senior Social Researcher, Department of Primary Industries)

We're seeing the consolidation of properties, therefore the role of towns as support centres for agriculture has dropped off. Farmers are operating with much larger ... more sophisticated machinery, and the local dealerships and mechanics are just not in a position to deal with that. Also some of the social connection has been lost from farming communities. As the farming profession has become more complex, they are relying not so much on, 'let's have a yarn over the fence,' it's more 'how do we get professional advice'? So what was a natural sharing of information and learning, we've lost some of that. So someone made the comment recently, 'the only time that farmers see each other is at clearing sales and funerals'. (Community Development Officer, Buloke Shire Council)

The changing nature of the dryland farm is having significant flow-on effects on farming families and communities. Increasing farm size and sophistication diminishes farmers' ability to work together and be serviced by local agribusiness. As a result, sharing and learning between farmers and local business is tempered. The leader of a local community group explained: 'there is the farming community and there's the town community, but one can't be without the other. The town community relies on the farming community, for business ... and the farmers rely on the community for services and social connectedness' (President, 'Ouyen Inc').

Finally, the impacts of agricultural and farming sector change are combined with broader, long-running, rural socio-demographic shifts that, in small settlements, have amplified effects. Donald is characterised by the more extreme trends of rural demographic change – a rapidly ageing and declining population base, the servicing of retirees and low socio-economic groups, and acutely limited employment and educational opportunities (Buloke Shire Council 2008). These shifts are introducing potential problems that, even without compounding pressures of drought and agricultural change, would threaten the socio-economic health of the smaller towns in the region. A sense of foreboding infused the way interviewees spoke about demographic change and its effects on Donald's prospects:

We're seeing an ageing and shrinking population. A lot of our farmers are much older now, they've sent their children away to be educated. A lot of those younger ones have enjoyed the city life and not come back. A number of them have wanted to come back, and the parents have said, 'don't do it to yourself, if you can't farm and make a dollar, why would you? Once you start losing people and services, there's the impact on schools ... and they can lose the capacity to host their own sporting teams, which is often the glue that holds communities together. So you can really see a difference in communities where they're losing this capacity. (Community Development Officer, Buloke Shire Council)

Donald, like other cropping towns, has been on the slow decline for generations. You've also had welfare migration and retiree migration. So the town population has been changing in those places, and what the drought's done, it's sped up that change ... they call it 'dust change'. And we're going to have significant service delivery challenges, as these people that are moving there don't have the assets backing them that the farm community had. They're often socially disadvantaged ... and don't understand the costs or strategies of living there. This should be seen as a failure of policy and planning, and drought just exacerbates that. I can't see these pressures changing ... it's long-term and much more reliable than drought. (Senior Social Researcher, Department of Primary Industries)

The challenges presented by this shifting rural landscape are many, ongoing and, perhaps, 'more reliable than drought' (Senior Social Researcher, Department of Primary Industries). Community groups, together with local councils, are actively working to ensure the sustainability of these rural towns through attempts to attract people and workers, maintain social connections and community, and promote place identity and amenity. However, all spoke of the difficulties presented by a declining population, skills base and pool of expertise, tired and 'burnt out' volunteers, and the ever-limited and geographically stretched funding and resources of small rural towns.

In these contexts, drought and a drying climate are just two of many threats. Long-term planning to manage these rural changes and to support associated transitions has generally lagged behind the difficult experience of change. It is clear from both Donald and Mildura that any such planning needs to address 'more-than-drought' and, to be effective, will need to do so in a coordinated and strategic way.

Understanding Drought-and-More in Context

Mildura and Donald are experiencing rapid change to their rural traditions and agricultural foundations – and these changes are common to many drought-affected farming areas. For the irrigated areas of Mildura, farmers are experiencing the challenges of learning to live in a drying environment – as part of a changing water industry and expanding global commodity market. In comparison, drought conditions are largely a normal part of agricultural production in the dryland regions of Donald. Drought in these areas has, in many ways, accelerated already occurring changes in both the farming industry and the demographic make-up of these small rural towns.

Economically, many farms are experiencing dwindling financial returns, assets and reserves after many years of drought and low commodity prices – and this is forecast to continue even with the break in drought conditions (RFCS 2008). Decreasing cash surplus and everyday spending (which also impacts on local town businesses), and ongoing and accumulating debts (RFCS 2008) are the combined effect. Compounding this are the extensive social impacts created by the 'drought-and-more' conditions confronting these regions. Service providers are now acknowledging the mounting social issues as pressures of drought (and subsequent floods), together with changing agricultural markets and commodity prices, continue to distress the community as a whole. Uncertainty is abounding in farming communities, producing stagnation and impeding change, adaptation and activity. In the context of enduring drying, such social impacts can become chronic, exhausting the resources of farming communities to foster community strength and well-being in the face of further long-term climatic and farming industry changes. For those living and working in Mildura and Donald, the aggregation of these issues threatens the well-being of farming communities.

Fundamentally, the issues facing both regions are 'more-than-drought', and any effective means of providing support to communities through these and future changes will require a holistic approach. As residents and representatives of Mildura and Donald continually emphasised, there is no one answer or panacea to the complex issues confronting these rural towns. The mode of crisis support available to farming communities to assist them through drought cannot address this complexity. Long-term, integrated planning for 'drought-and-more' needs to be developed.

Conclusion

Prior to this study, three reviews were carried out looking into the nature of drought and the government's response: an economic assessment of drought support measures by the Productivity Commission (2009); an assessment of the social impacts of drought on farm families and rural communities (the Drought Policy Review Expert Social Panel 2008); and a climatic assessment by the CSIRO and Bureau of Meteorology of the likely future climate patterns and the current EC standard of a one-in-20-to-25-year event (Hennessy *et al.* 2008). Each suggested that the government programmes used to support an adaptive response needed to affirm that prolonged periods of drought are natural and routine as opposed to an unexpected event, which was how governments traditionally conceived of such climatic events. Therefore, it is also necessary to ensure that any decision-making on drought response is undertaken independently of extreme drought events when public emotions and political effects are heightened. Similarly, drought adaptation strategies should not be shelved during periods of above average rainfall. Drought and flood adaptation strategies need to coexist – one should not replace the other as the climate oscillates between its wet and dry phases. This coexistence of strategies is especially important given the anthropogenic climate change projections for Australia, which suggest that increases in the frequency and duration of droughts will be associated with increases in the frequency of short-lived, but intense rainfall events (i.e. the type of weather that leads to flooding) (IPCC 2007, Whetton *et al.* 2005).

Likewise, it is important that the media recognise drought impacts take many forms and reporting needs to accurately reflect this. Communicating the reality of drought presents challenges for the media. As Wahlquist (2003) suggests, most reporters 'are city-based, with little conception of the complexities of the experience of drought' and 'radio and television weather presenters largely define good weather as the absence of rain'. Therefore issues such as increases in food prices due to drought shortages, environmental problems associated with increased drying and the loss of farming communities, do not register with city dwellers. It is only when issues like water restrictions kick in, that city-based media begin to take cognisance of prolonged dry periods. Likewise, when rains do reappear, media reportage tends to assume conditions are resolved. As with government drought assistance programs, the flawed logic fails to appreciate that drought is one element in a complex web of environmental, economic and socio-demographic changes that are mutually reinforcing.

Conversely, residents and representatives in Donald and Mildura have a keen awareness of the impacts of drought on their lives and work. By acknowledging the likely declines in water availability into the future, how this might play out as part of changing rural contexts, and how the coming together of these issues may fundamentally and permanently change the agricultural, farming and rural traditions of the region, we begin to understand the implications of a drought-prone climate. The issues farmers and farming communities face are not just a product of drought – to understand them as such would underestimate the extent of the problems and inhibit the ability to coordinate the holistic, cross-agency approach needed to address them.

Complex challenges come together and are experienced in very real and situated ways. This orientation provides a means of connecting broad rural economic, socio-demographic and climate trends to the everyday experiences of those living at the forefront of drought. The regions discussed in this chapter present a complex picture of drought-affected rural towns in Australia more generally.

References

Alston, M. and Kent, J., 2004. 'Coping with a Crisis: Human Services in Times of Drought', *Rural Society*, 14: 214-227.

Argent, N., 2005, 'The Neoliberal Seduction: Governing-at-a-distance, Community Development and the Battle over Regional Financial Service Provision in Australia', *Geographical Research*, 43: 29-39.

Argent, N., Smailes, P., and Griffin, A., 2007, 'The Amenity Complex: Towards a Framework for Analysing and Predicting the Emergence of a Multifunctional Countryside in Australia', *Geographical Research*, 45: 217-32.

Argent, N. and Walmsley, J., 2008, 'Rural Youth Migration Trends in Australia: An Overview of Recent Trends and Two Inland Case Studies', *Geographical Research*, 46: 139-152.

Australian Bureau of Agricultural Resource Economics (ABARE), 2008, *Australian Farm Survey Results 2005-2006 to 2007-2008*. [Online: Canberra, Commonwealth Government]. Available at: http://www.abareconomics.com/ publications.html/economy/economy.08/fsr-08.pdf [accessed: 10 September 2010].

Australian Bureau of Statistics (ABS), 2009a, *Australian National Accounts: National Income, Expenditure and Product*, Canberra: Commonwealth Government.

Australian Bureau of Statistics (ABS), 2009b, *Labour Force, Australia, Quarterly*, Canberra: Commonwealth Government.

Barr, N., 2005, *Understanding Rural Victoria*, Victoria: Department of Primary Industries.

Baum, S., O'Connor, K. and Stimson, R., 2005, 'Commentary says the Bush is in Bad Shape: Is that Really the Case?', *Fault Lines Exposed*, Melbourne: Monash University ePress, 06.1-06.39.

Birchip Cropping Group, 2008, *Critical Breaking Point? The Effects of Drought and Other Pressures on Farming Families*, Wimmera-Mallee: Birchip Cropping Group.

Botterill, L., 2009, 'The Role of Agrarian Sentiment in Australian Rural Policy', in F. Merlan and D. Raftery (eds), *Tracking Rural Change: Community, Policy and Technology in Australia, New Zealand and Europe*, Canberra: ANU E Press, 59-78.

Buloke Shire Council, 2008, *Community Profile*. Buloke Shire 2001 and 2006 census information.

Bureau of Meteorology (BoM), 2010, *Drought Statement Archive: for the 12-month period ending 31st December 2010*. [Online: National Climate Centre, ACT]. Available at: http://www.bom.gov.au/climate/drought/archive/20110107.shtml [accessed: 2 February 2011].

Bureau of Rural Sciences (BRS), 2008, *Social Atlas of Rural and Regional Australia Country Matters*, Canberra: Commonwealth Government.

Cocklin, C. and Dibden, J., 2005, *Sustainability and Change in Rural Australia*, Sydney: University of New South Wales Press.

Commonwealth Scientific and Industrial Research Organisation (CSIRO) and Bureau of Meteorology (BoM), 2007, *Climate Change in Australia: Technical Report*, Canberra: CSIRO Publishing.

Davies, A., 2009, 'Understanding Local Leadership in Building the Capacity of Rural Communities in Australia', *Geographical Research*, 47: 380-389.

Department of Primary Industries (DPI), 2010, *Food Business*, Melbourne: Victorian State Government.

Drought Policy Review Expert Social Panel, 2008, *It's About people: Changing Perspectives on Dryness*, Canberra: Commonwealth Government.

Golding, B. and Campbell, C., 2009, 'Learning to be Drier in the Southern Murray-Darling Basin: Setting the Scene for this Research Volume', *Australian Journal of Adult Learning*, 49: 423-450.

Gunasekera, D., Kim, Y., Tulloh, C. and Ford, M., 2007, 'Climate Change: Impacts on Australian Agriculture', *Australian Commodities*, 14: 657-676.

Hennessy, K., Fawcett, R., Kirono, D., Mpelasokaa, F., Jones, D., Bathols, J., Whetton, P., Stafford Smith, M., Howden, M., Mitchell, C. and Plummer, N., 2008, *An Assessment of the Impact of Climate Change on the Nature and Frequency of Exceptional Climatic Events*, Canberra: Bureau of Meteorology and CSIRO.

Intergovernmental Panel on Climate Change (IPCC), 2007, *Climate Change 2007: The Physical Science Basis, Summary for Policy Makers*, Cambridge: Cambridge University Press.

Mallee Catchment Management Authority, 2009, *Drought Impact: Irrigation Status Report for the Sunraysia Pumped Irrigation Districts 2008-09, Final Report*, Victoria: Mallee Catchment Management Authority.

McGuirk, P.M. and Argent, N., 2011, 'Population Growth and Change: Implications for Australia's Cities and Regions', *Geographical Research*, 49: 317-335.

Mildura Development Corporation (MDC), 2009, *Mildura Region Economic Profile*, Mildura, Victoria: Mildura Development Corporation.

Murray-Darling Basin Authority (MDBA), 2011, *Proposed Basin Plan*, Canberra: Commonwealth Government.

Pritchard, B., Burch, D. and Lawrence, G., 2007, 'Neither "Family" nor "Corporate" Farming: Australian Tomato Growers as Farm Family Entrepreneurs', *Journal of Rural Studies*, 23: 75-87.

Productivity Commission, 2009, *Government Drought Support Inquiry Report*, Canberra: Commonwealth Government.

Quiggin, J., 2008, 'Managing the Murray-Darling Basin: Some Implications for Climate Change Policy', *Economic Papers*, 27: 160-166.

Rural Financial Counselling Service (RFCS), 2008, *Public Submission to Drought Policy Review*, Submission No. 30 May, Canberra: Commonwealth Government.

Sherval, M. and Askew, L.A., 2012, 'Experiencing "Drought and More": Local Responses from Rural Victoria, Australia', *Population and Environment*, 33: 347-364.

Tonts, M. and Haslam-McKenzie, F., 2005, 'Neoliberalism and Changing Regional Policy in Australia', *International Planning Studies*, 10: 183-200.

Verdon-Kidd, D.C. and Kiem, A.S., 2009, 'Nature and Causes of Protracted Droughts in Southeast Australia – Comparison between the Federation, WWII and Big Dry Droughts', *Geophysical Research Letters*, 36.

Victorian Government, 2010, *Our Water, Our Future*. Available at: http://www. ourwater.vic.gov.au [accessed: 27 January 2011].

Wahlquist, A., 2003, *Media and Public Perceptions of Drought. Bureau of Rural Sciences* Seminar 13 June, 2003. Available at: http://asawahlquist.com/index2. php?option=com_content&do_pdf=1&id=11 [accessed: 23 May 2012].

Wei, Y., Langford, J., Willett, I.R., Barlow, S. and Lyle, C., 2011, 'Is Irrigated Agriculture in the Murray-Darling Basin well Prepared to Deal with Reductions in Water Availability?', *Global Environmental Change*, 21: 906-916.

Wheeler, S., Bjornlund, H., Zuo, A. and Edwards, J., 2012, 'Handing Down the Farm? The Increasing Uncertainty of Irrigated Farm Succession in Australia', *Journal of Rural Studies*, 28: 266-275.

Whetton, P.H., McInnes, K.L., Jones, R.N., Hennessy, K.J., Suppiah, R., Page, C.M., Barthols, J. and Durack, P.J., 2005, *Australian Climate Change Projections for Impact Assessment and Policy Application: A Review*, Canberra: CSIRO.

Wilson, G., 2009, 'The Spatiality of Multifunctional Agriculture: A Human Geography Perspective', *Geoforum*, 40: 269-80.

Winebiz, 2010, *Australia's Wine Industry Portal: Worldwide Comparisons*. Available at: http://www.winebiz.com.au/statistics/exports.asp [accessed: 10 June 2009].

Chapter 13

Evolving Metabolic Relations: Nature, Resources and People in the Hunter Valley

Phil McManus

Introduction

The recent emergence of 'patchwork' regions is a discourse intended to highlight the heterogeneity of Australia's regions. It is linked to the notion of a 'patchwork' economy, itself a concept designed to deflect attention away from the idea of a dual economy or two-speed economy that had the mining-oriented states of Queensland and Western Australia moving ahead of the other Australian states. Despite being more nuanced than the concept of a dual economy, the patchwork economy metaphor is limited and does not recognise the extent of diversity within rural and regional Australia, nor does it deal adequately with the causes of this diversity and the likely consequences of variation between and within regions.

The chapter begins with an overview and critique of the 'patchwork economy' metaphor. It then offers a brief introduction to the Hunter Valley before drawing on three major critiques of the patchwork economy metaphor to explore the Hunter Valley and its regional offspring, the Hunter Region, New South Wales. The chapter emphasises those conflicting industries and processes that are seen to contribute most significantly to this dynamic and complex region, namely coal mining, thoroughbred breeding, viticulture/ gastronomic tourism and demographic change (which is related to processes of land supply, housing costs and urban-rural migration). This exploration of various industries highlights the importance of legislation on mining, and the infrastructure capacity of the Port of Newcastle and the associated railway network, in influencing the future of other economic activities in the Hunter Valley, and beyond to the Liverpool Plains. The chapter then considers the evolving metabolic relations that emanate from, transverse and impact on the Hunter Valley. These relations are dynamic, and derive from the prevailing mode of production, in particular at the regional scale but always in relation to the destination of material flow in, through and from the region. Clark and Foster (2009: 314), observed that 'each mode of production generates a particular social metabolic order that influences the society-nature relationship, regulating the ongoing production of society and the demands placed on ecosystems'. While there are important metabolic relations involving thoroughbred breeding, viticulture and the production of gastronomic landscapes, and urban development, in this chapter it is argued that the most crucial metabolic relations influencing the Hunter Valley and the future of other industries and of many settlements are those associated with the mining, transportation and export of coal. Finally, the chapter elaborates on what this may mean for the sustainability of the Hunter Valley and beyond.

Patchwork Regions

The concept of Australia being a 'patchwork economy' was introduced by the Prime Minister of Australia, Julia Gillard, in October 2010 at a speech to the Queensland Media Club (Howells 2010). While being relevant to all of Australia, it was seen as having particular relevance within Queensland, which at the state level was perceived to be booming due to mining activity. In May 2011 the Treasury released a report titled *Investing in Australia's Regions*, noting that there are 'different opportunities and pressures right across Australia, with the nation's vast size and diverse patchwork economy presenting challenges' (Commonwealth of Australia 2011: 5). In the following paragraph, it was posited that 'the Government's role is to ensure that each unique part of the patchwork can develop to its full potential' (Commonwealth of Australia 2011: 5). In the space of one page, the Treasury shifted the scale of analysis from the state to the regional scale, constructed Australia's economy as being 'patchwork' and represented this spatially by showing different regions as internally homogenous.

The 'patchwork economy' metaphor relates to economic sectors, but when linked with regional development is spatially associated with particular regions within the national economy and various states. The notion of 'patchwork' is useful in reconceptualising economic growth away from a dualistic 'two-speed' or 'winner-loser' approach at crude levels of analysis, and for recognising that when applied to regions the individual identity of regions is retained, as opposed to the use of a 'melting pot' metaphor that creates uniformity. The patchwork metaphor is not new, being used previously in relation to cultural diversity (Forrest 1984, Chatterton 1999, Kurthen and Heisler 2009), political fragmentation (Fonchingong 2005), diversity in landscape and ecological zones (Ripple *et al.* 1991, Clark *et al.* 2010) and to describe an inadequate and probably temporary situation involving overlapping 'solutions' to a problem (Wolfe 1998, Steinemann 2004). When applied in the Australian regional and economic context, there are, however, three important weaknesses with the 'patchwork' metaphor. First, when applied at a regional scale it does not recognise variations and conflicts within the region. Second, it does not identify and interrogate the processes that create these variations. The processes that create a 'patchwork economy' and a 'patchwork of regions' also create variation within each region, and traverse regional boundaries. Third, and consequently, the metaphor does not interrogate the sustainability of the processes that are generating the 'patchwork' both in the aggregate picture of the region and within the region itself. These weaknesses are highlighted through the application of the 'patchwork' notion to the Hunter Valley.

The Hunter Valley

The Hunter Valley is defined as a water catchment of the Hunter River, which rises in the Barrington Tops and generally flows south-west before it is joined by the Goulburn River south of Denman and then flows south-east to enter the Pacific Ocean at what is now the city of Newcastle (See Figure 13.1). The catchment is approximately 21,500 square kilometres in area and extends further inland than any other coastal catchment in NSW (Australian Government 2009). Given the size of this river and its importance in the history of New South Wales, the valley has become the basis for the construct of the Hunter Region, an administrative unit that varies in boundary depending on the organisation adopting this unit. It is important, however, not to take a region as given, but to recognise that spaces and places are constructed both materially and discursively, and each modality of the constructions affects the others (Allen *et al.* 1998, McManus 2008a, 2008b).

Viticulture, coal mining and thoroughbred breeding are very important industries in the Hunter Valley. A number of conflicts have occurred between various industries competing for water, space, air quality, identity and labour, most notably in the 1990s between the coal mining industry and viticulture over the establishment of the Bengalla coal mine near Muswellbrook, which commenced operations in 1999 (McManus 2008a). The urbanisation of the coastal strip of land from Sydney north to Nelson Bay, including the Central Coast, Newcastle and Maitland, is increasingly placing pressures on water supplies and creating land use conflicts with agriculture and other activities. In 2006, during the drought, the NSW Premier announced a $342 million plan to construct the proposed Tillegra Dam to assist in 'drought-proofing' the Hunter and Central Coast region. The rejection in 2010 of the proposed Tillegra Dam meant that 4,500 hectares of fertile agricultural land in the Dungog Valley was not flooded (Sherval and Greenwood 2012).

Population growth was also the catalyst for the NSW government in 1997 to create two regions; the Lower Hunter Region, part of the Greater Metropolitan Region which stretches from Kiama to Nelson Bay, and the Upper Hunter Region, which is the area north of Singleton (McManus 2008b). In line with the demographic changes of population stagnation and decline in more remote rural areas, local government boundaries within the Upper Hunter Region were restructured in 2004, with the new Upper Hunter Shire being created out of

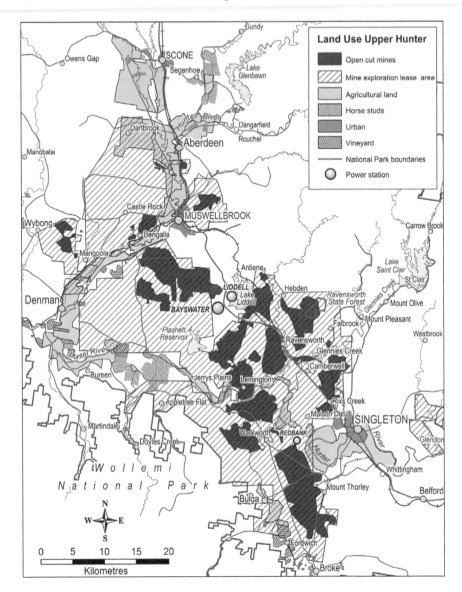

Figure 13.1   The Hunter Valley, NSW, Australia

Source: Satellite imagery Landsat Tm5, December 2011

the former local government area of Scone, and the eastern parts of the former Murrurundi and Merriwa shire councils. These current local government areas also highlight the diversity within the region, partly arising from their geographic location relative to such resources as commercially viable coal deposits.

Catalysts for Diversity

Coal Mining

Coal mining has a long history in Australia, and for many years the Hunter Valley, which forms part of the Sydney-Gunnedah Coal Basin, has been part of this history because of its extensive and commercially viable coal deposits. As the coalfields around Newcastle, at the mouth of the Hunter River, were 'mined out' in the late nineteenth and early twentieth century, allowing for what Wilson (1968) identified as 'scavenging' activity, the coal mining industry migrated up the Hunter Valley, from the Newcastle coalfield to the South Maitland coal field (Cessnock, Aberdare, Abermain, Richmond Main) and later to the Upper Hunter around Singleton, Muswellbrook and beyond the valley to Gunnedah (Wilson 1968, Day 1988, Productivity Commission 1998, McManus 2008a, 2008b). In the early twentieth century, the deep shafts around Cessnock became the most productive coalfields (Daly and Brown 1966), black coal fields where in 1954-1955 the 80 underground coal mines and six open cut coal mines in the region north of Newcastle were responsible for 9.4 million tonnes of raw coal production, mostly for domestic consumption (Gibson 1990). While Australia had been 'a viable international exporter before the First World War, the industry slid to the status of one of the most backward in the capitalist world during the inter-war period' (Gibson 1984: 227), a situation blamed on 'the general industrial turmoil in the industry' by the Productivity Commission (1998: C1). Export of coal was largely based on the use of coal as a paying form of ballast for the return voyage, and to power steamships, with the major destinations for Newcastle coal being San Francisco and Valparaiso in Chile (Jeans 1972). When these markets declined so did exports, until the rise of Japanese demand for coking coal in 1959 resulted in exports from the Bowen Basin in Queensland, followed by thermal coal exports to Asia (mostly Japan) in the 1980s (Productivity Commission 1998). In the Hunter Valley, by the 1960s, the Cessnock coalfields were old compared with international competitors and the resource was in decline. New, mainly open-cut, mines began operations near Singleton and later around the former dairy farming town of Muswellbrook (Day 1988). These coal fields supported new power stations at Liddell (completed 1974) and Bayswater (completed in the mid 1980s), because 'as recently as 1981, Australia was but a minor medalist in the global coal trade, exporting less than half as much coal as the US' (Pearse 2010: 22-23). Four years later Australia had become the world's largest exporter of black coal, a position that it still maintains (Productivity Commission 1998, Williams 2012).

While Australia produces only 6 percent of the world's coal, far behind China's production of 44 percent of world coal (Williams 2012), Australia was the world's leader exporter of coal in 2010, with 298 million tonnes, or 31 percent of the world's total exports (Australian Coal Association 2011a). Australia is a rare example of a country that exports most of the coal it mines (Cook 2003). Coal is transported by rail from the inland coalfields to major coal loading terminals on the coast, including Newcastle which has three terminals, with a fourth proposed terminal currently undergoing an environmental assessment. Newcastle handles more than 1000 coal vessels per year and is the largest coal export port in the world (Australian Coal Association 2011b). In the 2010-2011 Financial Year, coal exports exceeded 108 million tonnes, up from a record 97 million tonnes in 2009-2010. Japan, South Korea and Taiwan combined account for approximately 81 percent of coal exports, down from previous years due to increased exports to China (McManus 2008a, 2008b, Hunter Valley Coal Chain 2011). The expansion continues at a rapid pace, with the opening of the third coal terminal by Newcastle Coal Infrastructure Group in 2010 (with immediate approval for further expansion) creating extra capacity and enabling a record 121.9 million tonnes of coal to be exported in the 2011-2012 financial year (Boyle 2010, Newcastle Port Corporation 2012a).

The increased capacity of the port to export coal has resulted in bottlenecks further up the Hunter Valley Coal Chain (HVCC), the complex and rapidly growing network of mining, rail and port operations that is spread over 450 kilometres and 'is driven by the need to export very large coal volumes from a very concentrated network' (Hunter Valley Coal Chain 2011). The HVCC comprises 37 coal mines, 16 producers, 29 trains (generating 15,000 trips per year), and three Coal Loading Terminals at the mouth of the Hunter River containing a total of seven ship berths (Hunter Valley Coal Chain 2011, EMGA/MitchellMcLennan 2012), with expansion plans currently being assessed for environmental impacts. Importantly, the Environmental Impact Statement for Port Waratah Coal Services constructs the proposed port expansion as being required to 'accommodate projected and contracted future coal exports' (EMGA/MitchellMcLennan 2012: 1). Conversely, added capacity at the port creates additional pressures on rural areas, remnant bush land and other land uses in the Upper Hunter and surrounding areas where the coal is mined.

In essence, the NSW Mining Act 1992 allows any person to apply for a mining lease (Section 51) and Section 68 includes provisions that allow that a mining lease may be granted over land of any title or tenure, in respect of any mineral or minerals regardless of the nature of ownership and 'over the surface of land and the subsoil below the surface, over the surface of land and the subsoil down to a specified depth below the surface or over the subsoil between or below any specified depth or depths below the surface of land' (NSW Government 1992, Section 68). This legislation places all other activities in the Hunter Valley and elsewhere on the defensive in relation to activities such as coal mining, regardless of their economic importance, long-term sustainability and contribution to the identity of the region. The legislation enables what Karl Marx called 'primitive accumulation' which involves 'taking land, say, enclosing it, and expelling a resident population to create a landless proletariat, and then releasing the land into the privatised mainstream of capital accumulation' (Harvey 2003: 149). Such accumulation is, understandably, resisted by vested interests in the Hunter Valley whose metabolic relations differs from those of the coal mining industry, but are inextricably linked to this powerful industry given competition for resources.

One important and powerful group of vested interests in the Upper Hunter Region is the thoroughbred horse industry. Their power and influence was shown in 2010 with the rejection of the proposed Bickham Coal Mine near the Pages River, upstream of the most valuable thoroughbred breeding farms in Australia (Connor *et al.* 2008, Munro 2012, McManus *et al.* 2012, 2013). The anticipated lifespan of the proposed mine was 22 years, with an annual output of 2.5 megatonnes, making it small compared to other mines in the region (NSW Department of Planning 2005, Connor *et al.* 2008, Munro 2012). Even though the mine was to be small on the scale of open-cut mining in the Upper Hunter region, it was clearly a contest for land, water, transport and regional identity. If the Bickham proposal had been accepted, it would have become the first operating mine in the Upper Hunter Shire Council area, despite the existence of coal mining in the past and the fact that coal mines exist to the south, west and north of this local government area. The influence of the thoroughbred breeders in the Upper Hunter through the formation and support of the Bickham Coal Action Group is similar to the situation further to the north-west of this region, where long-term landowners in the fertile and agriculturally productive Liverpool Plains area are fighting coal mining proposals by BHP-Billiton and the Chinese company Shenhua and coal seam gas extraction proposals by Santos (Munro 2012).

Coal seam gas (CSG) exploration and drilling operations have also had a significant impact in the Hunter Region. Gases are trapped, along with water, in coal deposits deep underground. They can be extracted using various methods, with most concern about hydraulic fracturing (fracking) where chemicals and water are pumped into shale beds or coal seams to cause cracking and release the gas. There is opposition to CSG operations on prime agricultural land in the Hunter Valley, where there are already approximately 35 exploration wells, with a concentration of activity around Singleton. Nearby to the east at Gloucester there are 42 exploration wells within 20 kilometres of the town and north on the Liverpool Plains there are numerous wells around Narrabri. These areas highlight the concern that despite its world renowned thoroughbred breeding and wine production industries, the Hunter Valley could become a focus for CSG activity, resulting in the destruction of environments that support other industries.

Thoroughbred Breeding

Only a few places around the world are closely associated with thoroughbred breeding. Thoroughbred breeding in the Upper Hunter region centres on the town of Scone, approximately 250 kilometres north of Sydney, which had a population of 5079 at the 2011 census and is indisputably the 'horse capital of Australia' (Australian Bureau of Statistics 2011). There are approximately 65 studs in the region, from which over 70 percent of Australia's thoroughbred foals are conceived, and which includes all of Australia's top sires and broodmare (NSW Department

Figure 13.2   Upper Hunter Horses and Powerlines

of State and Regional Development 2008). These thoroughbred breeding farms are a source of cultural identity for the town of Scone and the Upper Hunter region (see McManus 2008a, 2008b, McManus *et al.* 2011). This cultural identity is built on a long history of thoroughbred breeding in the region, boosted by recent developments in thoroughbred racing, particularly by the Scone Race Club.

The Upper Hunter region emerged as a thoroughbred breeding region in the 19th century. The earliest horse breeding in the Hunter Valley developed from horses bred at Windsor, north-west of Sydney, that were imported into the Hunter Valley (Guilford 1985, White 2005). By the 1830s a number of settlers who were interested in horse breeding established studs in the Hunter Valley because their previous locations at Windsor and Prospect were 'considered inferior to the "new country" in the immense watershed of the Hunter Valley' (Walden 2004: 1). The number of stallions in the Hunter more than trebled between 1830 and 1840, and Sydney owners began to send their stallions to the Hunter Valley for all or part of the breeding season (Guilford 1985). This provided a basis for the thoroughbred breeding industry. The first race meeting in the Hunter Valley held was in Maitland in 1833, followed by Scone in the 1840s (Guilford 1985).

The racing side of the industry developed relatively slowly. The Scone Race Club opened a new racetrack in 1994 and the significant prize money allocated to the Emirates Park Scone Cup, plus the establishment of the lucrative Inglis Scone Guineas, perpetuate the thoroughbred association with Scone and the Upper Hunter Region (Thomas 2011, McManus *et al.* 2013). The growing importance of thoroughbred racing in Scone reinforces the equine identity of Scone and the Upper Hunter region, and strengthens its position relative to competing industries in the region.

Viticulture and Gastronomic Tourism

The rural areas of the Hunter Valley vary from the broadacre agriculture of the western part of the valley, where wheat, sheep and cattle dominate, through to the more intensive agricultural pursuits of the Lower Hunter. Activities such as thoroughbred breeding are not classified as agriculture by the Australian Bureau of Statistics, and serve the needs of urban visitors, visibly when people attend stallion days, stud tours or when thoroughbred owners arrange to visit particular stud farms. It is definitely the situation for the viticulture industry, which involves cellar-door sales, wine tasting, music concerts 'in the vines' and close links to accommodation facilities and recreational activities such as golf and hot air ballooning (Gibson and Connell 2012, O'Neill and Whatmore 2000).

The viticulture industry of the Lower Hunter (particularly around Pokolbin and Broke) differs from the Upper Hunter (mainly around Denman) because of its proximity to Sydney (McManus 2008a). Such wine regions are part of the 'functional expansion of the city' (McManus 2005: 85), despite wine tourism having an obvious connection with the rural landscape. This is recognised by promoters of Hunter Valley wine tourism who note that 'locals have done everything to create the perfect escape' by providing '60 restaurants, 120 wineries, 160 accommodation venues and a vast range of activities', in a region that is 'less than two hours drive from Sydney' (Hunter Valley Wine Industry Association n.d). The different location and foci of the wine producing areas has meant that the branding and tourism promotion of these areas has been local, rather than regional. The stronger sub-regional associations (such as Hunter Wine Country and Port Stephens Tourism Association) were intent on promoting 'local tourism clusters' based on wines or beaches and dolphins, rather than promoting the Hunter as a region (Dredge 2005). Despite the growth in wine tourism, vineyards in the Lower Hunter region were the category most likely to change land use patterns, both gaining from other land uses and losing land to other activities (Manandhar *et al.* 2010). This was primarily a result of land value increases in the region between 1985 and 2005, with urban development (particularly in Pokolbin but to a lesser extent in Rothbury and Cessnock) displacing vineyards. The loss of vineyards to urban development is a manifestation of urban expansion pressures emanating from Sydney and coastal areas near Newcastle, as part of wider demographic changes highlighted below. Simultaneously, loss of grape production areas within vineyards has occurred due to the relative success of music concerts, with some vineyards pulling up vines to expand concert venues. Bimbadgen Estate, with an amphitheatre capable of accommodating 8,000 people, and Hope Estate with their massive

19,000 seat amphitheatre, demonstrate the success of Hunter Valley wineries in creating facilities for music events that promote the brand and associate it with desirable experiences (Gibson and Connell 2012, see also Chapter 10).

Demographic Change

Industrial development and urbanisation pressures emanating from the expansion of Sydney and other coastal centres are contributing to an internal patchwork within the Hunter Valley. In essence four key demographic changes are taking place. First, coastal population growth is extending inland to more affordable settlements, such as Maitland, which is increasing in population. This coastal growth is putting pressure on available water supplies, and was a motivating factor for the proposed, but abandoned, Tillegra Dam (Sherval and Greenwood 2012). The growth of Maitland and Singleton, and the expansion of coal mining have been major reasons for the loss of 744,755 hectares, or 43 percent, of land used for food production in the Hunter Statistical Division between 1980 and 2010 (Kelly 2011). A spokesperson from the NSW Minerals Council disputed this analysis, claiming that 'land used by mining and waste industries in the Hunter was estimated at 20,500 hectares, 0.6 percent in 2006' (Kelly 2011). The transition from dairy and beef farming to thoroughbred breeding is probably another factor in this loss of land for food production, although thoroughbred breeding operations are presented as rural activity (McManus *et al.* 2011, 2013). Second, urban-rural migration is contributing to the growth of desirably located towns and the surrounding semi-rural areas (Connell and McManus 2011). This is occurring in Scone, where its horse image and the lack of coal mining operations is encouraging population growth, including residents who work in the mining industry (McManus 2008b). In 2006, 151 men (some 12 percent of employed male residents) in the urban locality of Scone worked in the mining industry, compared with just 9 women (1 percent of female residents in employment) (ABS 2006, see also Chapter 7). The town of Scone has expanded into former agricultural land and abuts thoroughbred breeding farms, including historic St. Aubins stud to the south of the town. The changing demography is highlighted by the approved expansion of the town onto part of the former St. Aubins Estate to meet the future residential, aged care and educational needs of the town. While urban expansion around Scone has resulted in the loss of some land formerly used by thoroughbred breeders, the scale of thoroughbred breeding in the area has increased as these farms have taken over land formerly used for beef and dairy cattle production. Third, more remote areas such as Merriwa and Murrurundi have been declining and ageing in population, although there is some turnaround in Murrurundi as cheap housing in the town and the growth of Gunnedah and areas to the north improve the strategic location of the town (see Table 13.1). Fourth, former villages that were not of sufficient size to be considered urban localities by the ABS, and hence do not appear in the tables derived from Census data in this chapter, have been severely impacted by coal mining activity. Bylong, Camberwell, Ulan and Wollar are examples of

Table 13.1 Demographic Change in the Hunter Valley, 1991-2011

Urban Locality/Year	1991	1996	2001	2006	2011
Aberdeen	1797	1737	1708	1791	1837
Cessnock-Bellbird	17,932	17,540	17,791	18,316	20,013
Denman	1509	1440	1400	1385	1403
Kurri Kurri-Weston	13,268	12,555	12,317	12,532	13,057
Maitland	45,209	50,108	53,391	61,431	67,132
Merriwa	962	937	993	946	973
Murrurundi	983	902	783	805	847
Muswellbrook	10,140	10,541	10,010	10,222	11,042
Paterson	312	347	341	345	372
Scone	3329	3468	4555	4624	5079
Singleton	11,861	12,519	12,495	13,665	13,961

Sources: ABS (1998, 2006, 2011)

former small rural settlements that have lost population or been eradicated due to the expansion of coal mining activity. Success within the patchwork economy is accompanied by negative consequences for some small rural communities in the region.

Metabolic Relations

The Hunter Valley, compared with many parts of Australia, is export oriented. As shown in Figure 13.1, metabolic flows enter the region through the provision of manufactured products (machinery, electronic goods, clothing) imported from overseas and the importation of resources such as petrol and diesel. These are combined with domestic resources such as sunlight, water, grass, oats and coal to provide the total natural resources input (including possible recycling within imported products). Increased use of recycling within the Hunter Valley (for example, recycled water being used for dust suppression in coal mines) is included to account for the total resource input. Importantly, this analysis of metabolic flows highlights the dependency on the environment and the traversing of regional boundaries. Outputs from the region include the export of coal, thoroughbreds, wine, alumina and electricity. Some of the metabolic flows remain in the region, as 'accumulation' in the building of houses, roads and railway lines, or consumed

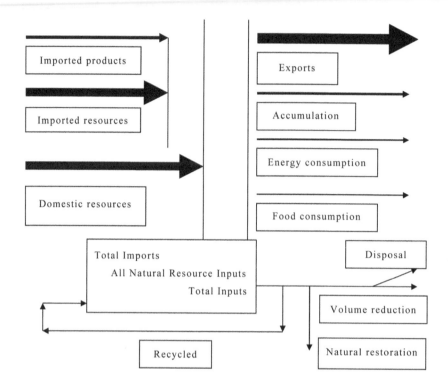

Figure 13.3   Indicative Metabolic Flows In and Through the Hunter Valley

Source: Developed from Government of Japan (2008: 22)

in the form of energy or food. Some materials are disposed of as waste, while others are used to restore the environment (water, replanting of vegetation). As shown in Figure 13.3, the recycling of materials, including the industrial recycling of fly ash for use in cement by power stations, with about 5 percent of coal ash being recycled by Bayswater and Liddell power stations in 2003 (Brown *et al.* 2006), and the domestic recycling of paper, glass and 'green waste' is part of the metabolic cycle. Within the region, the availability of water that is 'fit for purpose' is a crucial metabolic flow. This means that potable water must be available for urban settlements for drinking, cooking and other functions where health and safety are paramount, while recycled water circulates through the system and can be used for dust suppression and for other purposes where potable water is not necessary.

The actual size of the flows can be measured in various ways, with differing results. For example, if the indicative flows shown in Figure 13.3 were measured variously by weight, volume or dollar value, the thickness of the arrows would change, because the price of coal and other commodities fluctuates. Figure 13.4 shows the flow of one commodity, coal, in and through the Hunter Valley, from being mined through to being consumed as electricity in Australia or exported to Japan, South Korea and Taiwan, for thermal coal (to generate electricity) or as coking coal for use in steel production and therefore emitting carbon. Hunter Valley coal is important for the generation of electricity in Australia (with

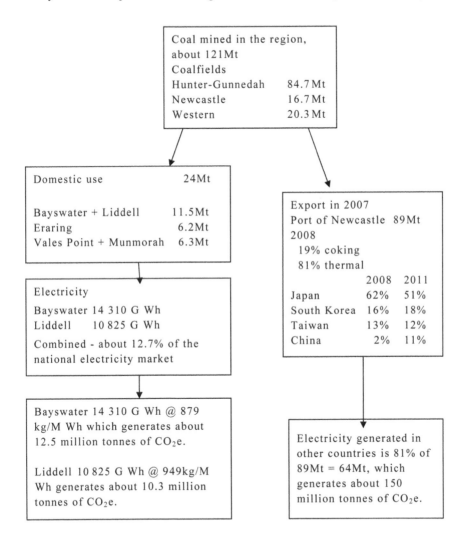

Figure 13.4   Coal Mining, Use and Impacts, 2007

Sources: Brown *et al.* 2006, Port Waratah Coal Services 2008, Australian Coal Association 2011a, 2011b, Newcastle Port Corporation 2012a

Bayswater and Liddell combined comprising 13 percent of the National Electricity Market in 2007) while Eraring, Vales Point and Munmorah power stations on Lake Macquarie also contribute substantially to this market. Hunter Valley coal is vital for Australia's exports, with Port Waratah Coal Services (operator of the Carrington coal terminal, one of the two Kooragang Island coal terminals and the proposed T4 coal terminal in Newcastle) being 18 percent owned by Japanese trading companies, 7 percent owned by Japanese steel companies and 5 percent owned by Japanese power, cement and oil companies (EMGA/MitchellMcLennan 2012). The Hunter Valley and the port of Newcastle are important nodes in networks that connect Muswellbrook with Tokyo and Singleton with Seoul. How sustainable are these networks, both for the specific metabolic relations of the network and for the earth when the metabolic relations involving the Hunter Valley are viewed from a planetary perspective?

Sustainability

The sustainability of the environment is dependent on the continuation of natural systems that support life, including photosynthesis, hydrological cycles and soil formation processes. In rural areas, this means the avoidance of unnecessary clearing of vegetation (particularly in areas vulnerable to erosion) and the maintenance of top soil by crop rotation and the avoidance of over-stocking. In the Hunter Valley, the expansion of the coal industry and the threat of coal seam gas exploration using methods such as fracking is a major threat to sustainability because they disrupt and destroy the natural systems that support life *in situ*. The coal industry eradicates the soil profile, despite efforts at rehabilitation, while coal seam gas production poisons water bodies and disrupts the flow of water (Munro 2012). For these reasons there is great concern in the Hunter Valley, and north-west on the Liverpool Plains, about the incompatibility of coal mining and agriculture. While the NSW Minerals Council claims that 'the vast majority of mining takes place on lower-value grazing land or land that isn't arable' (quoted in Kelly 2011), the expansion of both coal and coal seam gas exploration and mining into the Hunter Valley and onto the fertile Liverpool Plains (where grain and seed yields are 40 percent above the national average) is being challenged by residents who can trace their ownership of rural properties to the 1820s and 1840s, making them 'landed gentry in Australian terms' (Munro 2012: 188). The extent of opposition to coal seam gas production was seen in a protest outside the NSW Parliament House in Sydney in early May, 2012, when often disparate organisations such as The Greens, the Country Women's Association and numerous local activist groups came together to rally against the NSW government's proposals for handling this issue. As the President of the Country Women's Association (CWA), Elaine Armstrong, told the rally: 'This is more than just tea and scones ... This is about protecting our land and water'. Equally telling was the fact that this was the first time that the CWA had marched on Sydney in its 90 years of existence.

The issue of sustainability is not, however, confined to the water catchment boundaries of the Hunter Valley, or to the administrative boundaries of the Hunter Region. While there are multiple cross-boundary sustainability issues to be considered, the most important of these is the generation of Greenhouse Gas emissions through the burning of thermal coal mined in the Hunter Valley (see Figure 13.4). Drawing on mining data, statistics on the domestic use of coal, breakdowns of coking and thermal coal exports, the calorific value of coal as outlined by the Commonwealth of Australia (2011) and calculations for similar sustainability issues for the China First mine proposal in the Galilee Basin in Queensland (Rutovitz and Mason 2011), it is conservatively calculated that in 2007, the domestic and international burning of coal mined in the Hunter Valley contributed about 212 million tonnes of Carbon Dioxide equivalent (CO_2e) per annum to the atmosphere. This figure does not appear in Australia's carbon accounting because, as shown by Rutovitz and Mason (2011), 96 percent of the total emissions of the proposed China First mine were generated when the coal was burned, with the remaining 4 percent being generated by mine construction, mine operations, rail transport to the port and international shipping of coal from Queensland to China. In the case of the Hunter Valley, this means that most emissions are accounted for by Japan (to which 51 percent of all types of coal exports were destined from Newcastle in 2011). Since 2007, coal exports from Newcastle have increased from 89 million tonnes per annum (Mtpa) to almost 122 Mtpa in 2011-2012 (Newcastle Port Corporation 2012a). Projected exports of coal from Newcastle vary from 180 Mtpa in 2015 (Boyle 2010) to the even larger figures of 220 Mtpa in 2014, 244 Mtpa in 2019 and 278 Mtpa in 2024 (ACIL Tasman 2009). The Newcastle Port Corporation (2012b 2) believes that while coal export capacity at the port will rise to 330 Mtpa if the T4 proposal is approved as submitted, coal exports 'are unlikely to exceed 275 Mtpa between 2017 and 2025' due to infrastructure constraints on the railway system. As a result of the Newcastle Infrastructure Group terminal opening and then expanding, in 2016 the terminal contracted capacity at the Port of Newcastle will be 243 Mtpa, but contemporary coal chain capacity is *only* (my emphasis) 170 Mtpa (Newcastle Port Corporation 2012b: 2). Additional capacity enabling rising coal exports means additional pressure on other land uses in the Hunter Valley and beyond to the Gunnedah Basin as well as an increase in carbon emissions for the planet.

At a time when the amount of carbon dioxide in the atmosphere is approaching 400 parts per million (Davis *et al.* 2010, Earth System Research Laboratory, Global Monitoring Division 2012), a figure never seen before in the human occupation of the planet, and the links between carbon dioxide and other greenhouse gases and climate change are compelling, the addition of more carbon through the burning of coal in both Australia and internationally raises serious questions about sustainability at a planetary level. While the Hunter Valley will not be most impacted by a warming scenario (compared with the deltaic megacities in parts of Asia), the long-term effects of climate change are likely to see reduced rainfall over the catchments of existing dams and industries, such as thoroughbred breeding, move south to cooler locations where rainfall is more predictable. The sustainability issue extends beyond the spatial and temporal boundaries of the Hunter Valley today, while actions taken today are likely to impact the future for many years to come.

Conclusion

The Hunter Valley is typical of rural areas in other parts of Australia with rural-urban migration coinciding with the ageing and declining of populations in more remote agriculture areas within the region. It is also typical of urban overspill in the growth of cities such as Maitland, and in the popularity of 'playgrounds' such as the Pokolbin and Broke vineyard regions of the Lower Hunter close to Sydney (with similar regions to be found near Melbourne, Adelaide and Perth). The Hunter Valley is similar to the Bowen, Surat and Galilee coal basins in Queensland, where large inland deposits of coal are mined in vast open cut mining operations, the coal is railed to the port and shipped to export markets. Coal from the Hunter Valley goes mainly to Asia, and increasingly to China which has increased its share of coal exported from Newcastle from 2 percent in 2008 to 11 percent in 2011 (Port Waratah Coal Services 2008, 2011). The trajectory of thoroughbred breeding is similar to that in the Yarra Valley near Melbourne where fertile land with good water, but close to the major metropolitan market, has been overtaken by urbanisation pressures so that the main breeding region in Victoria is now in north-east Victoria.

What is unusual is the juxtaposition of these mutiple trends and industries in one geographic location. While these activities are often spatially separated, as seen with the Bengalla Coal Mine and the rejected proposal for the Bickham Coal Mine, conflict occurs over land, water, access and the identity of the region. These conflicts are intensifying, particularly as CSG adds another land use impact to an environment currently perceived by some residents as already degraded by extensive coal mining operations. The Hunter region is a physical boundary (as in a water catchment) that has become constructed as a region. This region is internally diverse, and is traversed by many conflicting processes and metabolic flows that are shaped by, and help shape, the region. This internal diversity and the importance of transboundary flows are overlooked and understated in the patchwork economy metaphor.

Significantly, the notion of a patchwork economy places no emphasis on sustainability. A region may appear to be a 'winner' in the patchwork economy of Australia, but the consequences for sustainability may be negative. In the case of coal mining in the Upper Hunter and its export to countries such as Japan, South Korea, Taiwan and China for burning in power stations, climate change impacts are likely to be experienced mostly by poor people in low-lying parts of Asia. This need not necessarily be the case. Improving the sustainability of the Hunter Valley by not approving any new coal mines, by developing renewable energy schemes, implementing a 'just transition' of mining workers and communities and fostering more sustainable agricultural activities within this region, will assist to improve the sustainability of the planet. This will necessitate a transition strategy in countries that are currently importing Australian coal, and will assist countries such as India to avoid becoming dependent on a non-renewable fossil fuel that is contributing to the warming of the earth. The sustainability of the Hunter Region and the planet means that it is more important to focus on the metabolic relations involving the Hunter Valley than to try and improve the relative position of this region within the patchwork economy.

Acknowledgements

Thank you to Laurence Troy for his help with the preparation of Figure 13.1 and to the editors for their helpful feedback on various versions of this chapter.

References

ACIL Tasman, 2009, *Vision 20/20 Project: The Australian Minerals Industry's Infrastructure Path to Prosperity* (an assessment of industrial and community infrastructure in major resources regions), Melbourne: ACIL Tasman Pty Ltd.

Allen, J., Massey, D., Cochrane, C. with Charlesworth, J., 1998, *Rethinking the Region*, New York: Routledge.

Australian Bureau of Statistics, 1998, *1996 Census of Population and Housing. Selected Characteristics for Urban Centres and Localities*, New South Wales and Australian Capital Territory, Canberra: Australian Bureau of Statistics.

Australian Bureau of Statistics, 2006, *Census Home*, available at: http://www.censusdata.abs.gov.au [accessed 19 July, 2012].

Australian Bureau of Statistics, 2011, *Census Home*, available at: http://www.abs.gov.au/websitedbs/censushome.nsf/home/data?opendocument#from-banner=LN [accessed 3 December, 2012].

Australian Coal Association, 2011a, *The Australian Coal Industry – Coal Exports*, available at: http://www.australiancoal.com.au/facts-and-figures.html. [accessed: 7 February, 2012].

Australian Coal Association, 2011b, *The Australian Coal Industry – Coal Loading Ports*, available at: http://www.australiancoal.com.au/ports-and-transport. html. [accessed 7 February, 2012].

Australian Government, 2009, *Australian Natural Resource Atlas*, available at http://www.anra.gov.au/topics/water/overview/nsw/swma-hunter-river-unregulated.html [accessed: 20 April, 2012].

Boyle, B., 2010, *Report 3: NSW Coal Export Forecast to 2015: Forging Alignment of Mines, Rail and Port Infrastructure*, AustCoal Consulting Alliance. PDF

available at: http://austcoalconsulting.com/component/option,com_frontpage/
Itemid,1/ [Accessed 13 February, 2012].

Brown, P., Cottrell, A., Searles, M., Wibberley, L. and Scaife, P., 2006, 'A Life
Cycle Assessment of the New South Wales Electricity Grid (year ending 2003)',
Pullenvale, Queensland: Technology Assessment Report 58. Cooperative
Research Centre for Coal in Sustainable Development.

Chatterton, P., 1999, 'University Students and City Centres – The Formation
of Exclusive Geographies. The Case of Bristol, UK', *Geoforum*, 30: 117-
133.

Clark, R.W., Brown, W.S., Stechert, R. and Zamudio, K.R., 2010, 'Roads,
Interrupted Dispersal, and Genetic Diversity in Timber Rattlesnakes',
Conservation Biology, 24: 1059-1069.

Clarke, B. and Foster, J.B., 2009, 'Ecological Imperialism and the Global Metabolic
Rift: Unequal Exchange and the Guano/Nitrates Trade', *International Journal
of Comparative Sociology*, 50: 311-334.

Commonwealth of Australia, 2011, *National Greenhouse Accounts Factors*,
Canberra: Department of Climate Change and Energy Efficiency.

Connell, J. and McManus, P., 2011, *Rural Revival? Place Marketing, Tree Change
and Regional Migration in Australia*, Farnham: Ashgate Publishing.

Connor, L., Higginbotham, N., Freeman, S. and Albrecht, G., 2008, 'Watercourses
and Discourses: Coalmining in the Upper Hunter Valley, New South Wales',
Oceania, 78: 76-90.

Cook, B., 2003, *Introduction to Australia's Minerals: Coal (Coal in a Sustainable
Future)*, Adelaide: Osmond Earth Sciences.

Daly, M. and Brown, J., 1966, *The Hunter Valley Region, NSW*, Newcastle: The
Hunter Valley Research Foundation.

Davis, S.J., Caldeira, K. and Matthews, H.D., 2010, 'Future CO_2 Emissions and
Climate Change from Existing Energy Infrastructure', *Science*, 5997 (329),
1330-1333.

Day, D., 1988, 'Evolutionary or Fragmented Environmental Policy Making?
Coal, Power and Agriculture in the Hunter Valley, Australia', *Environmental
Management*,12: 297-310.

Dredge, D., 2005, 'Local versus State-Driven Production of "The Region":
Regional Tourism Policy in the Hunter, New South Wales, Australia', in A.
Rainnie and M. Grobbelaar (eds), *New Regionalism in Australia*, Aldershot:
Ashgate Publishing, 301-319.

Earth System Research Laboratory, Global Monitoring Division, 2012, Trends in
Atmospheric Carbon Dioxide. Online: Available at http://www.esrl.noaa.gov/
gmd/ccgg/trends/#mlo_full [Accessed 4 May, 2012].

EMGA/MitchellMcLennan, 2012, T4 Project: Environmental Assessment
prepared for Port Waratah Coal Services Limited, February 2012, Sydney:
EMGA/MitchellMcLennan.

Fonchingong, C.C., 2005, 'Exploring the Politics of Identity and Ethnicity in State
Reconstruction in Cameroon', *Social Identities*, 11: 363-380.

Forrest, L., 1984, 'A Patchwork of People (Surinam)', *Geographical Magazine*, 56: 82-89.

Gibson, C. and Connell, J., 2012, *Music Festivals and Regional Development in Australia*, Farnham: Ashgate Publishing.

Gibson, K., 1984, 'Industrial Reorganisation and Coal Production in Australia 1860-1982: An Historical Materialist Analysis', *Australian Geographical Studies*, 22: 221-242.

Gibson, K., 1990, 'Internationalization and the Spatial Restructuring of Black Coal Production in Australia', in R. Hayter and P.D. Wilde (eds), *Industrial Transformation and Challenge in Australia and Canada*, Ottawa: Carleton University Press, 159-173.

Government of Japan, 2008, *Fundamental Plan for Establishing a Sound Material-Cycle Society*, Tokyo: Government of Japan.

Guilford, E., 1985, 'The Glendon Stud of Robert and Helenus Scott, and the Beginning of the Thoroughbred Breeding Industry in the Hunter Valley', *Journal of Hunter Valley History*, 1(1): 63-106.

Harvey, D., 2003, *The New Imperialism*, Oxford: Oxford University Press.

Howells, M., 2010, 'Queensland Developing Patchwork Economy, says Gillard', ABC News, [Online 12 October, 2010]. Available at: http://www.abc.net.au/news/2010-10-12/queensland-developing-patchwork-economy-says/2295052 [Accessed 30 March, 2012].

Hunter Valley Coal Chain, 2011, *About Us*, available at: http://www.hvccc.com.au/AboutUs/Pages/MapOfOperations.aspx [accessed 28 February, 2012].

Hunter Valley Wine Industry Association, no date, *Hunter Valley Uncorked*, available at http://www.huntervalleyuncorked.com.au/visitor-info [accessed 5 June, 2012].

Hunter Valley Wine Industry Association and the Hunter Valley Protection Alliance, 2012, *Protecting the Hunter Valley from CSG Mining*, Hunter Valley Wine Industry Association and the Hunter Valley Protection Alliance.

Jeans, D.N., 1972, *An Historical Geography of New South Wales to 1901*, Sydney: Reed Education.

Kurthen, H. and Heisler, B.S., 2009, 'Immigrant Integration: Comparative Evidence from the United States and Germany', *Ethnic and Racial Studies*, 32: 139-170.

Kelly, M., 2011, 'Coal Industry Eating up Hunter Food Bowl', *Newcastle Herald*. 14 October, Online: 14 October, 2011. Available at: http://www.theherald.com.au/news/local/news/general/coal-industry-eating-up-hunter-food-bowl/2323366.aspx?src=rss [Accessed 16 May, 2012].

McManus, P., 2005, *Vortex Cities to Sustainable Cities: Australia's Urban Challenge*, Sydney: UNSW Press.

McManus, P., 2008a, 'Their Grass is Green, but Ours is Sweeter: Thoroughbred Breeding and Water Management in the Upper Hunter Region of New South Wales, Australia', *Geoforum*, 39: 1296-1307.

McManus, P., 2008b, 'Mines, Wines and Thoroughbreds: Towards Regional Sustainability in the Upper Hunter, Australia', *Regional Studies*, 42: 1275-1290.

McManus, P., Albrecht, G. and Graham, R., 2011, 'Constructing Thoroughbred Breeding Landscapes: Manufactured Idylls in the Upper Hunter Region of Australia', in S.D. Brunn (ed.), *Engineering Earth: The Impacts of Megaengineering Projects*, Dordrecht: Springer, 1323-1339.

McManus, P., Albrecht, G. and Graham, R., 2013, *The Global Horseracing Industry: Social, Economic, Environmental and Ethical Perspectives*, Abingdon: Routledge.

Manandhar, R., Odeh, I.O.A. and Pontius Jr., R.G., 2011, 'Analysis of Twenty Years of Categorical Land Transitions in the Lower Hunter of New South Wales, Australia', *Agriculture, Ecosystems and Environment*, 135: 336-346.

Munro, S., 2012, *Rich Land, Wasteland: How Coal is Killing Australia*, Sydney: Pan Macmillan.

Newcastle Port Corporation, 2012a, *Trade Statistics*, available at: http://www.newportcorp.com.au/site/index.cfm?display=111694 [accessed 3 December, 2012].

Newcastle Port Corporation, 2012b, *Coal Exports through Port of Newcastle will Not Exceed 275Mtpa Defore 2025*, Newcastle: Newcastle Port Corporation.

New South Wales (NSW) Department of Planning, 2005, *Coal Mining Potential in the Upper Hunter Valley – Strategic Assessment*, Sydney: NSW Department of Planning.

NSW Department of State and Regional Development, 2008, *Hunter Means Business: Investment Prospectus 2008*, Sydney: NSW Department of State and Regional Development.

New South Wales Government, 1992, *Mining Act*, 1992. Act 29 of 1992. As of 4 April, 2012, available at: http://www.austlii.edu.au/au/legis/nsw/consol_act/ma199281/ [accessed 3 May, 2012].

O'Neill, P. and Whatmore, S., 2000, 'The Business of Place: Networks of Property, Partnership and Produce', *Geoforum*, 31: 121-136.

Pearse, G., 2010, 'King Coal', *The Monthly*, May, 20-26.

Productivity Commission, 1998, *The Australian Black Coal Industry: Inquiry Report Volume 2: Appendices*, Canberra: Productivity Commission.

Ripple, W.J., Bradshaw, G.A. and Spies, T.A., 1991, 'Measuring Forest Landscape Patterns in the Cascade Range of Oregon, USA', *Biological Conservation*, 57: 73-88.

Rutovitz, J. and Mason, L., 2011, *Climate Impact Analysis of the China First Mine and the Proposed Development of the Galilee Basin, Queensland*, Sydney: UTS Institute for Sustainable Futures.

Sherval, M. and Greenwood, G., 2012, '"Drought-proofing" Regional Australia and the Rhetoric Surrounding Tillegra Dam, NSW', *Australian Geographer*, 43: 253-272.

Steinemann, A., 2004, 'Human Exposure, Health Hazards, and Environmental Regulations', *Environmental Impact Assessment Review*, 24: 695-710.

Thomas, R., 2011, 'Scone Racing Club Receives Quality Nominations for Cup Carnival', *The Daily Telegraph*, available at: http://www.dailytelegraph.com.au/sport/superracing/scone-racing-club-receives-quality-nominations-for-cup-carnival/story-fn67r1j3-1226052814102 [accessed 5 December, 2011],

Walden, H., 2004, *The Spirit Within: Scone's Racing History*, Sydney: Harley Walden.

White, J., 2005, *Horses in the Hunter, 1820-2005*, Scone: The Seven Press.

Williams, N., 2012, The CCS Experience in Australia. Speech to the Annual Clean Coal Forum, Beijing, 30 March.

Wilson, M.G.A., 1968, 'Changing Patterns of Pit Location on the NSW Coalfields', *Annals of the Association of American Geographers*, 58: 78-90.

Wolfe, J.M., 1998, 'Canadian Housing Policy in the Nineties', *Housing Studies*, 13: 121-133.

Index